Artificial Intelligence in Managing Antimicrobial Resistance

Editors

Ramendra Pati Pandey
School of Health Sciences and Technology
UPES University, Dehradun, India

Chung-Ming Chang
Department of Medical Biotechnology and Laboratory Science
Chang Gung University, Taiwan (R.O.C.)

V. Samuel Raj
Director, Centre for Drug Design Discovery & Development (C4D)
Dean Academic Affairs, SRM University, India

CRC Press
Taylor & Francis Group
Boca Raton London New York

CRC Press is an imprint of the
Taylor & Francis Group, an **informa** business
A SCIENCE PUBLISHERS BOOK

First edition published 2025
by CRC Press
2385 NW Executive Center Drive, Suite 320, Boca Raton FL 33431

and by CRC Press
4 Park Square, Milton Park, Abingdon, Oxon, OX14 4RN

CRC Press is an imprint of Taylor & Francis Group, LLC

Library of Congress Cataloging-in-Publication Data (applied for)

ISBN: 978-1-032-81245-8 (hbk)
ISBN: 978-1-032-82112-2 (pbk)
ISBN: 978-1-003-50301-9 (ebk)

DOI: 10.1201/9781003503019

Typeset in Time New Roman
by Prime Publishing Services

Preface

Antimicrobial Resistance (AMR) presents a growing global threat, challenging healthcare systems worldwide and jeopardizing the gains made in medicine over the last century. As pathogens evolve to resist available treatments, we find ourselves on the brink of a crisis where simple infections could once again become untreatable. Addressing this challenge requires innovative approaches that go beyond traditional methods of antibiotic development and infection control.

Artificial Intelligence (AI) has emerged as a transformative technology that offers immense potential in addressing the complexities of AMR. By leveraging AI's capabilities in data analysis, predictive modeling, drug discovery, and personalized medicine, we have an unprecedented opportunity to combat this crisis in new and effective ways. AI's role extends beyond the laboratory, offering tools for precision medicine, diagnostic advancements, and even supporting the global effort of antimicrobial stewardship.

This book chapter explores how AI is being integrated into the fight against AMR, highlighting its role in understanding antimicrobial mechanisms, predicting antibiotic resistance, and discovering new therapeutics. We delve into various aspects such as AI-driven diagnostics, disease surveillance, and antimicrobial stewardship, while also addressing the limitations and ethical challenges that come with implementing AI in healthcare.

The discussion starts by examining the history and evolution of AI in combating AMR, followed by an in-depth exploration of its applications in drug design, predictive models, and optimization of treatments. Real-world case studies, such as AI in pediatric infectious disease prognosis and AI-driven protein structure prediction, are presented to illustrate the power of AI in bridging the gap between complex biological data and actionable healthcare solutions.

It is our hope that this chapter will provide a comprehensive overview of the ways AI can transform AMR management, inspiring researchers, clinicians, and policymakers to embrace technological innovations in their fight against one of the most pressing health challenges of our time. While the promise of AI is immense, challenges remain—ranging from ethical considerations to the need for collaborative global efforts. This chapter seeks not only to highlight AI's potential but also to initiate a broader dialogue on how best to harness this technology for sustainable solutions in AMR management.

Contents

1

Role of Artificial Intelligence in Fighting Antimicrobial Resistance

Ruby Dhiman, Chanchal, Prashant Anil Kumar Singh, Sitabja Mukherjee and *Ramendra Pati Pandey**

Antimicrobial resistance (AMR) is a critical global health threat driven by extensive antibiotic use. The chapter "Role of Artificial Intelligence in Fighting Antimicrobial Resistance" explores how AI can combat AMR. It traces AI's history in healthcare, from the Dendral project to IBM Watson, showcasing AI's evolving role in diagnosing and treating infections. The chapter delves into antimicrobial peptides (AMPs) and their dual mechanisms, highlighting their potential as alternatives to traditional antibiotics. It covers AI's applications in predicting pediatric infectious diseases, optimizing antibiotic administration, forecasting resistance, and aiding in protein structure prediction and drug design. Challenges such as health communication complexities, poverty's role, professional awareness gaps, and international data discrepancies are addressed, emphasizing the need for integrated strategies. Despite AI's potential, the chapter notes its limitations, including data quality issues and ethical considerations. In conclusion, AI's integration into healthcare can enhance AMR management by improving diagnosis, treatment, and

School of Health Science and Technology (SOHST), UPES, Dehradun – 248007, Uttrakhand, India.

* Corresponding author

drug development, offering a promising approach to controlling resistant infections.

1.1 Introduction

Antibiotic resistance has been brought on by the extensive use of antibiotics in clinical practice, which has also raised the risk of the formation of super-resistant bacteria. The current era of antibiotics began in 1928 with the discovery of penicillin by Alexander Fleming. Since then, countless people with bacterial and fungal diseases have been spared from death because of antibiotics. However, the extensive use of antibiotics in clinical practice has increased the risk of the creation of super-resistant bacteria and led to drug resistance [1]. Microorganisms that are resistant to many drugs, multiple drugs combined, or even pan-drug resistance are frequently seen in hospital settings, particularly in the critical care unit. The European Union/European Economic Area (EU/EEA) estimates that bacteria resistant to antibiotics cause around 670,000 illnesses annually, with an approximate 33,000 fatalities. Antimicrobial resistance is still widespread throughout the WHO European Region, especially in the southern and eastern regions. Patients infected with Klebsiella pneumoniae and Acinetobacter spp. have fewer options for antibiotics due to the bacteria' growing resistance to third-generation cephalosporins and carbapenems, as well as the fast spread of resistant isolates [2]. The discovery of antimicrobial drugs that are well-tolerated by the human body has led to the groundbreaking ability to slow or inhibit the development of harmful germs *in vivo*. The fatality of bacterial infections started to drastically decline in the early half of the 20th century, confirming the antimicrobials' revolutionary effect on human history.

Furthermore, by enabling the effective treatment of dangerous infections that arise following significant modifications to our defensive barriers, such as following major surgery or chemotherapy for malignancies, antimicrobials have significantly contributed to the success of other ground-breaking breakthroughs in contemporary medicine. However, the enormous benefits of antibiotics are being threatened by the development of antimicrobial resistance (AMR) [3]. With the advancement of technology in recent years, artificial intelligence (AI) has proven remarkably helpful in managing antimicrobial resistance (AMR), one of the top 10 worldwide risks to public health that people will encounter this century [4]. This suggests that, despite the current global crisis, we are making progress against drug-resistant microorganisms. Modern technology may be needed for the research and implementation of AI-based applications, but because of AI's automation and efficiency, the resulting products, algorithms, or processes are more accessible to end users, even in subpar healthcare settings, demonstrating AI's influence and equitable role in AMR control [5].

The accessibility of data from electronic health records (EHR) and advancements in processing efficiency have been key drivers of AI progress. These two elements have a tight relationship with intricate mathematical operations like neural networks (NN) and machine learning (ML) [6]. This has become even more pertinent with the development of deep neural network (DNN) designs, whose complexity (often defined as the quantity of parameters the networks must learn) has skyrocketed in the previous 10 years [7]. A subset of artificial intelligence called machine learning (ML) is different in that it can adapt to new information. While ML attempts to automatically learn rules, much like the human brain does, expert systems are manually designed based on human expertise. Unless in some restricted cases, ML does not require human participation [8]. On the other hand, neural networks (NNs) are mathematical informatics computation models that are based on how the biological brain networks, whether human or animal, functions. Put another way, NNs are models made up of information connections that have the ability to identify the underlying correlations in a dataset. A DNN is made up of several layers (typically more than five) of processing units that enable us to learn to grasp the data on our own and improve predictions. Neural networks have the notable benefit of gradually improving with dataset size [7], allowing them to adjust to changing inputs [9].

Patient management is complicated today since there are many factors pertaining to a patient's care and medical background. According to a recent article, 200 times more medical knowledge will be published this year than any person could ever read in their lifetime [10]. AI can significantly improve healthcare decision making by automatically handling this enormous volume of data. But most medical professionals still do not see the benefits of AI, and they continue to base their choices only on their own experiences and established protocols [11].

We can improve our capacity to stop the spread of antibiotic resistance, maximize treatment outcomes, and preserve the efficacy of antibiotic treatments for coming generations by incorporating AI-driven solutions into healthcare systems and public health campaigns.

1.2 History of AI in AMR

Stanford University academics created the first problem-solving software, "Dendral," in the 1960s with the intention of assessing hypotheses. Organic chemistry pioneers were to utilize it to identify unknown materials by analyzing their mass spectra. This initial method was used to determine which bacteria were causing severe blood infections and to suggest the best course of antibiotic treatment [12]. However, legal issues have made it impossible to employ this approach in clinical practice [13].

The application of AI in healthcare gained traction in 2016 when a 60-year-old woman's unusual form of leukemia was identified by AI software incorporated into the IBM Watson platform, which also suggested an efficient course of treatment [14]. One of the earliest papers on the application of AI in pediatrics was published in 1984, and it featured an automated health decision-making system called SHELP, which was intended to identify metabolic abnormalities that are inborn [15]. About thirty years later, the Boston Children's Hospital effectively leveraged the IBM Watson platform to offer helpful assistance in the diagnosis and treatment of uncommon pediatric disorders [16].

Publications about the application of AI in pediatrics have increased dramatically since 1984. These include emergency management (automatic risk assessment of appendicitis), among other areas [17], assistance in making diagnostic decisions [18], and as a framework for early hospital treatment prediction and asthma exacerbation [19]); pediatric oncology (gene expression profiling of children with neuroblastoma or lymphoblastic leukemia [20] and comparative study of important genes for the development of anticancer medicines [21]). as well as infantile neuropsychiatry (identifying motor abnormalities [22], predicting seizures in infants with epilepsy [23], and identifying children with autism spectrum disorder based on facial abnormalities [24])

1.3 Understanding Antimicrobial Mechanisms of Action

Microorganisms and humans alike contain tiny peptide molecules known as antimicrobial peptides (AMPs), a large family of compounds. They are the first line of defense against invading microorganisms because they can trigger an innate immune response in a variety of animals [25].

Since AMPs are thought to be potential antibacterial drugs, a great deal of research has been done on their mode of action. To improve the use of AMPs as medicinal agents, characterization is essential [26]. Compared to traditional antibiotics, antimicrobial peptides have less microbial resistance and have both bacteriostatic and bactericidal actions [27]. These peptides are amphiphilic molecules with a positive charge and hydrophilic and hydrophobic residues. After attaching to the cell membrane by a variety of methods, including the barrel stave model, the carpet model, the toroidal pore model, etc., these peptides finish their function [28]. The disruption of the plasma membrane, intracellular antimicrobial mechanism, inhibition of the synthesis of macromolecules like protein, nucleic acids, and enzyme activity, and antimicrobial effect by taking part in immune regulatory effects are some of the hypothesized mechanisms of action of these peptides [29,30].

Based on their secondary structure, AMPs may be classified into four main categories: linear α-helical peptides, β-sheet peptides, linear extended peptides, and peptides combining both α-helix and β-sheet [31]. Via the barrel-stave pore model or carpet model, the helical peptides cause damage to membranes. Their primary job is to introduce amphipathic helices into the membranes of bacteria, which causes the membranes to become disorganized [32]. The β-sheet peptides have several functions, such as inhibiting the development of cell walls and attaching to certain lipid elements in membranes [33].

AMPs' mode of action may be generally divided into two groups: direct killing and immunological control (Fig. 1). Two further groups of direct killing MOAs exist: membrane targeting and non-membrane targeting. Membrane permeabilizing peptides, which are mostly identified by cationic peptides like defensin, LL37, and melittin, have the ability to form temporary holes in membranes. Non-membrane targeting peptides such as pleurocidin, pyrrhocidin, and mersacidin can penetrate the cell membrane and disrupt essential cellular functions that eventually result in cell death without causing the membrane to become permeable [29].

1.3.1 Targeted killing: the action's membrane-targeting mechanism

AMPs disrupt negatively charged membranes by binding through hydrophobic and electrostatic interactions to lipids with anionic head groups found in bacterial outer membranes, such as cardiolipin and phosphatidylglycerol. Because of the positive charge on their α-helix surface, which is critical for killing germs, AMPs may interact with negatively charged microbe membranes and reveal their antimicrobial activity. The zwitterionic phospholipids in mammalian membranes are only weakly interacting with the hydrophobic portions of AMPs (Fig. 2). The binding of AMPs to mammalian cell membranes is reduced by cholesterol. The net charges, amphiphilic characteristics, and hydrophobicity of amino acid-based macromolecules (AMPs) determine their essential impact on the selective action against microorganisms [30]. The electrostatic interaction of negatively charged bacterial membranes with cationic AMPs and anionic phospholipids is essential for cellular communication. The hydrophobic interface between cationic AMPs and mammalian membranes is quite modest since the outer surface of eukaryotic cell membranes is composed of phospholipids with phosphatidylcholine head groups and sphingomyelin with a small amount of ganglioside. There is substantial interaction between the hydrophobic part of AMPs and the outer surface of bacterial membranes because negatively charged phospholipids are present [34].

There are two methods by which the membrane-targeting AMPs interact: receptor-mediated and non-receptor-mediated.

1.3.1.1 Mechanism mediated by receptors

This is facilitated by a limited set of AMPs, or receptor-mediated peptides, which are made up of pore-forming and receptor-binding domains. They typically function by interacting with membrane components to resist bacteria *in vitro* at micromolar or nanomolar concentrations [35].

Most AMPs produced by bacteria, viruses, and tumor cells have this mechanism; examples of these include nisin, lacto-coccin, and mesentericin [35]. A further illustration of a receptor-mediated process to the membrane receptor SbmA is PR-39. Proline-arginine-rich and linear in form, PR-39 is a cathelicidin AMP [32]. Although this AMP cannot form the pores in the bacterial membrane, it is known to have a variety of multifunctional properties, including the ability to heal wounds by inhibiting the expression of syndecan, to reduce inflammation by inhibiting NADPH oxidase, to attract neutrophils and leucocytes through chemoattraction, and to intervene in protein and DNA synthesis by quickly inducing proteolytic activity, which causes some proteins involved in DNA replication to degenerate [32,36].

1.3.1.2 Mechanism mediated by non-receptors

The majority of vertebrate and invertebrate AMPs that exert their activity by interacting with membrane components are mostly part of the non-receptor mediated action mechanism [26]. Gram-positive bacteria have teichoic acid on their outer membrane, whereas Gram-negative bacteria have lipopolysaccharide. Both substances cause a net negative charge on the membrane surface and bind to cationic AMPs via electrostatic attraction [37,38]. The most studied mechanism to comprehend the MOA of AMPs is membrane permeability. When AMPs attach to microbial membranes, they break down the membrane structures of bacteria or cancer cells, causing the contents of the cells to leak out and ultimately causing the cell to die [39,40,41].

There are two processes in the process of cell membrane damage. The negatively charged bacterial cell membranes are first preferentially bound by the cationic AMPs, which then rupture or fail to perforate the membranes. Four models—the barrel-stave model, the carpet model, the toroidal-pore model, and the aggregated channel model—are hypothetical models that fall under the membrane perforation mode. According to the non-perforation method, AMPs attach to the bacterial cell membranes and impede normal biological functions including protein synthesis, RNA transcription, and DNA replication to kill the cells [28].

1.3.2 Targeted killing: a method of action that targets non-membranes

AMPs that target the internal components of bacteria and those that target the cell wall of bacteria are the two main groups into which the non-membrane targeting MOA is generally divided [26].

1.3.2.1 Targeting peptides in the cell wall

Like conventional antibiotics, AMPs work by interacting with a range of precursor molecules that are necessary for the creation of cell walls to block the manufacture of new cell walls. Lipid II is one such precursor molecule that AMPs primarily target [42]. The mechanism of action of human α defensin 1 [37] and human β+ defensin 3 [43] is that they bind to lipid II, demonstrating their bactericidal properties.

1.3.2.2 Targeting intracellular components using peptides

AMPs have the ability to cross bacterial cell membranes and interact with intracellular targets including DNA and RNA, which can disrupt the physiological processes of the bacterium. This may interfere with the creation of proteins and cell walls [38]. Prior to targeting intracellular components through the inhibition of essential cellular activities, AMPs engage in interactions with the cytoplasmic membrane. Intracellular targets are involved in mechanisms that block the formation of cell walls, prevent the creation of macromolecules like proteins or nucleic acids, or stop enzyme activity. Certain AMPs, such as buforin II and indolicidin, can pass through and penetrate bacterial membranes, attach to nucleic acids (DNA or RNA), and therefore prevent the creation of new nucleic acids [26]. Since certain AMPs do not cause membrane permeabilization at the lowest possible dose yet still cause bacterial mortality, several of them now target intracellular components [38].

1.3.3 Mechanism of action of immunological regulation

AMPs have the ability to suppress the immune system in addition to directly targeting and eliminating germs [41]. Different mechanisms are used by AMPs to exhibit immune-modulatory effects, including the reduction of endotoxin-induced inflammatory response, the induction of pro-inflammatory proteins and cytokines, the regulation of adaptive immunity, and the recruitment of macrophages [44,45,46]. These peptides do not directly kill germs; instead, they strengthen the body's defenses against pathogens [47].

AMPs are a component of innate immunity and are sometimes referred to as "the first line of defense" because they are among the first molecules generated by immune cells such as neutrophils and macrophages to combat alien invaders [47]. Certain antigen-presenting cells (AMPs) exhibit a range of immunological responses, including the induction and development of white blood cells (WBCs), the inhibition of inflammatory chemokine production, and the regulation of chemokine and reactive oxygen/nitrogen species expression [48,49,50,51,52]. In mammals, AMPs stimulate the immune system in a number of ways, including (i) T cell activation; (ii) Toll-like receptor stimulation; and (iii) other mechanisms. Increased phagocytosis; (iv). Activation of dendritic cells; (v). The neutrophils' chemoattraction [53].

Numerous cells in the body, including phagocytes, keratinocytes, lymphocytes, epithelial cells, and neutrophils, create AMPs. These cells may be found in the gastrointestinal tract, immunological system, genitourinary tract, and lymphatic system.

Some immune cells, such as macrophages and neutrophils, make AMPs constitutively, whereas other immune cells, such as epithelial cells, produce them in response to stimulation of the mucosal surface [54]. The majority of β-defensins are formed via induction, whereas α-defensins are commonly synthesized as part of AMPs [55]. Human AMPs, such as LL-37 and β defensins, have the ability to draw in immune cells, including mast cells [44], dendritic cells, and leukocytes [56].

1.4 AI Against AMR

1.4.1 Pediatric infectious disease prognosis, evaluation, and diagnosis

Early detection of infectious pathology, the ability to distinguish between diseases with an infectious vs non-infectious foundation, and appropriate care of consequences are critical components of the battle against antibiotic resistance. Compared to adults, children have greater infection rates, and they frequently have non-specific symptoms, which makes diagnosis more difficult [57]. In this manner, AI has the potential to be a very formidable weapon.

A method based on reinforcement learning was presented by Komorowski et al. in 2017. It allows a virtual agent to learn a set of rules from a trial-and-error system in order to improve them and maximize the predicted performance [58]. Through the analysis of numerous clinical judgments, this technology was able to extract a volume of patient data that outweighed the life experience of human physicians and taught itself the best way to treat sepsis. Patients whose doctors' actual judgments aligned with AI's have seen reduced mortality as a result of its use, demonstrating the tool's clinically verifiable capacity to tailor sepsis therapy and support physicians in making decisions in real time [58].

An early diagnosis of infectious sepsis from non-infectious forms of SIRS was the goal of an experiment carried out in a German pediatric tertiary critical care unit. The theory behind this trial was that the two conditions share many symptoms [59]. In order to achieve this goal, a machine learning (ML) diagnostic model that utilized a random forest technique was created. This model considered 44 factors that were accessible at the time of patient admission, including baseline characteristics, clinical and laboratory measurements, and technological and medical assistance. The approach made it possible to identify all cases of sepsis at an early stage, and it was estimated that patients with non-infectious SIRS might potentially need 30% less antibiotics [59]. Liang et al. presented a follow-up pediatric research in 2019 that included an analysis of 101.6 million data points from 567,498 outpatients.

Fifty-five diagnostic codes representing common pediatric disorders were evaluated in the initial diagnosis. Acute tonsillitis, bronchitis, bronchopneumonia, and acute upper respiratory tract infections were among the most common diagnoses [60]. But the system also performed admirably when it came to diagnosing potentially fatal illnesses like meningitis. A hierarchical diagnostic system was established through the use of logistic regression classifiers in the analysis. This system performed exceptionally well in all organ systems and subsystems, exhibiting a high degree of accuracy between the expected diagnoses and the initial diagnoses made by a medical examiner [60].

Because ML-based systems can examine EHR in a manner similar to the fictitious deductive reasoning employed by doctors, they may be utilized for tasks like evaluating triage protocols or supporting doctors in the identification of difficult-to-treat diseases. The decrease in unnecessary testing and associated expenses is a pertinent benefit [61].

1.4.2 Sufficient administration of antibiotics

Generally speaking, prescribing the right antimicrobials is a difficult task because it entails choosing the right course of action for the suspected pathogen, controlling the concentration and frequency of the antimicrobial agent, and figuring out the best way to guarantee that the medication reaches the infection site [62]. In the pediatric domain, it is imperative to acknowledge that the nature of infection and resistance exhibit notable age-related variations, and the range of dosages based on age and weight is extensive [63].

One of the challenges associated with prescribing antibiotics is the requirement to gradually modify a patient's course of treatment in response to new clinical findings. Hospitals are increasingly depending on automated decision support systems to review antimicrobial prescriptions because manual surveillance is unsustainable due to a lack of specialized healthcare resources and the volume of data that needs to be processed. Most prescription monitoring systems use a rule base acquired from published and expert guidelines to identify inappropriate prescriptions and prevent potential adverse events. These systems are often poorly defined and therefore generate a high rate of clinically unhelpful alerts [61].

Systems built on machine learning have been developed to solve this issue. APSS, an antibiotic monitoring system, was introduced by Beaudoin et al. in 2014. Unlike earlier systems, APSS could pick up new prescription surveillance criteria. This functionality was created to allow APSS to eventually self-improve its knowledge base in conjunction with user input monitoring. The capacity of APSS to learn indications allowed for a suitable and clinically meaningful switch from intravenous to oral antimicrobial medication [64].

In a follow-up study, the learning module was able to extract clinically relevant rules by identifying inappropriate prescriptions not recognized by the baseline system, using the same APSS system to identify prescribing practices not supported by local antimicrobial stewardship experts [63].

In underdeveloped nations, the issue of rising health care expenses and a lack of personnel to properly prescribe antibiotics is especially critical. Because of this, a team of researchers postulated in 2018 that low-cost, personalized predictions for specific empirical antibiotic selections may be obtained by utilizing machine learning techniques to easily gather patient data [61].

Blood culture data collected from a 100-bed children's hospital in north-western Cambodia between February 2013 and January 2016 were analyzed, and information on clinical characteristics, demographic characteristics, and living conditions was acquired using 35 independent variables, processed via machine learning algorithms to predict Gram stain results and whether bacterial agents could be treated with common empirical antibiotic regimens. The results showed that modern ML algorithms can amplify widely used logistic regression models by predicting antibiotic susceptibility [65].

Furthermore, automated systems have a significant impact on the timely prescription of antibiotics by monitoring the side effects of these treatments in real time. By putting into practice and assessing an automated monitoring system that was altered to identify medication-associated adverse events in pediatric patients, including those linked to antibiotics, Kilbridge et al. demonstrated the potential of these techniques [66].

The best techniques based on AI and, more recently, ML, when integrated into a decision support system, could significantly enhance the proportion of patients receiving successful empirical antibiotic therapy while lowering the dangers associated with increasing resistance selection.

1.4.3 Resistance to antibiotics prediction

Predicting antibiotic resistance is one area where artificial intelligence is showing promise. Given that diagnostic tests and antibiotic resistance assays can take a long time, this means that AI can help doctors treat their patients more effectively. Furthermore, because age-related differences in susceptibility exist, empirical therapy in the juvenile setting is more complicated than in adult medicine [63]. Yelin et al. examined a 10-year longitudinal dataset containing over 700,000 UTIs that were acquired in the community. They found a high correlation between the resistance to six examined antibiotics and patient demographics, prior urine culture findings, and medication purchase history. After that, they created a customized machine learning model for predicting antibiotic resistance, which revealed that some drugs (such nitrofurantoin) had a greater peak risk during early life and infancy [67].

Another approach was created to forecast antibiotic resistance; however, it works via previously unidentified channels, employs genomically-based algorithms to forecast the bacterial phenotype, and is strengthened by the ongoing expansion of high-density genomic data sets for a variety of microbes. A bioinformatics tool called VAriant Mapping and Prediction of Antibiotic Resistance serves as an illustration of this use. It was created using machine learning techniques and was used to analyze sequencing data from 3393 bacterial isolates of nine species that had antibiotic resistance phenotypes for 29 different antibiotics. With an average accuracy of 91.1% for all antibiotic-pathogenic combinations, the researchers identified 14,615 variant genotypes and created 93 association and prediction models that validated the mechanisms of genetic resistance to recognized antibiotics [68].

Over 10,000 samples of M. tuberculosis from 16 countries across six continents have been tested using a comparable method, which has shown accurate in identifying the point mutations linked to the evolution of antibiotic resistance to first-line medicines [69].

1.4.4 AI in protein structure prediction and drug design

The inefficiencies inherent in the drug discovery process are often emphasized by increasing costs, extended timelines, and a high rate of failures. Few drug candidates advance to clinical trials, and many of them fail as late as Phase 3, meaning that the overall success rate in clinical drug development is only about 10–20% [70]. Estimates of the total costs associated with research and development before product launch range from $161 million to $4.54 billion in 2019 U.S. dollars per successful drug [71]. Among the difficulties in drug development is the lack of good quality structural models of potential therapeutic targets, which is one of AlphaFold's primary concerns. But as this research demonstrates, AI is also moving quickly to alleviate other drug development obstacles.

Prior to the development of AlphaFold, protein structure prediction often involved many phases, as the description that follows illustrates.

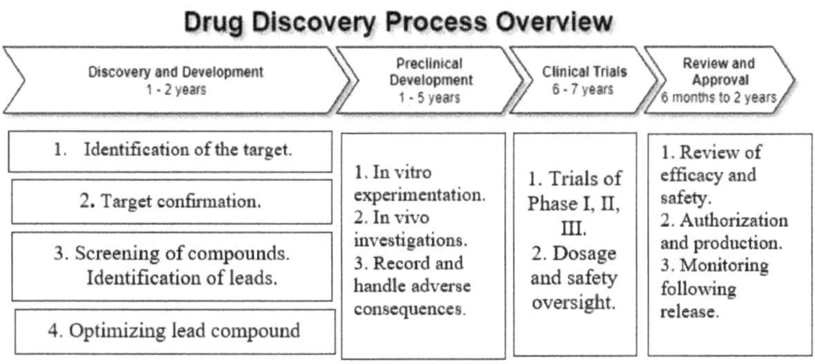

Figure 1: Stages of Drug Discovery Process.

1.4.5 Equivalency and comparative frameworks

Homology modeling uses the structure of a homologous protein to predict the three-dimensional structure of a protein. The process consists of four steps: target identification (finding a homologous protein with a known structure); alignment (matching the target sequence to the template); model building (building a model of the target protein from aligned regions); and refinement (improving the accuracy and stability of the model). The HHpred method and related set of tools represent advances in remote homology recognition and alignment between distant homologies [72,73]. This issue was tackled from a different perspective by predictions of protein contact maps derived from coevolution patterns [74], which were further strengthened by the initial uses of deep learning neural networks [75]. Tools like Rosetta [19] and I-Tasser [76] moved beyond homology and toward comparative modeling in the late 2010s [77]. Smaller components of known structures, along with an energy-like scoring function and empirical folding criteria, were used by Rosetta to accomplish this. Similar to I-TASSER (Iterative Threading ASSEmbly Refinement), this approach combines fragment assembly with template-based modeling. The similarity of other techniques has begun to approach that of ab-initio protein structure prediction [78]. These developments directly influenced the creation of AlphaFold and the subsequent AI methods for predicting protein structures.

1.4.6 Validation of structure

The accuracy and plausibility of the predictions are guaranteed via structure validation. The dihedral angles in the Ramachandran plot are verified and structural feature geometry is analyzed using tools such as [79]. Potential energy is measured via energy-based assessments, such ANOLEA, in order to determine if folding is proper. If accessible, predicted structures can then be compared to experimental data. Predictions for proteins without established experimental structures cannot be evaluated using such comparisons; instead, they can be used to benchmark the prediction algorithms and determine expected accuracy. Nevertheless, functional hypotheses derived from the anticipated three-dimensional structures, including the identification of active site or interaction interface residues, may be examined *in vitro*, thus providing an indirect validation of the structural prediction [80].

1.5 AI in Drug Repurposing

Because of its exceptional efficiency in reducing costs and time, governments and pharmaceutical corporations are beginning to pay more and more attention to the viable approach of drug repurposing. AI technology also aided in the transfer of drugs, saving time and money [81]. Figure 2 describes the medication repurposing workflow.

New computational techniques for identifying connections between many kinds of biological entities, including genes, parts, illnesses, and medications, can be advantageous for drug repurposing. The benefit of this research is that it can find new therapeutic uses for medications that are already on the market [81].

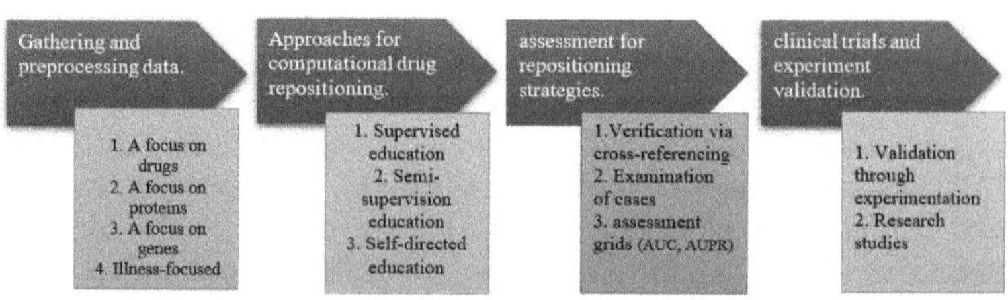

Figure 2: Drug Repurposing Workflow.

Chen et al. [82] addressed drug combination as an additional type of drug repurposing by building a sizable computational biology network made up of medications, genes, and illnesses to assess the interaction between target and biomolecule. Using a variety of machine learning techniques (random forest, SVM, and logistic regression), Kim et al. [83] explored possible relationships between medications and illnesses while accounting for self-defined similarity metrics (disease-drug and drug-drug similarities).

Large-scale data may be used by real-world research projects to accurately represent the course of diagnosis, therapy, and patient health in real-world settings. Deep learning's exceptional strength may be used to signal and mine riches from huge amounts of real-world data, while standard statistical approaches are restricted in their ability to handle large volumes of data. A methodology for drug repurposing was developed by Liu et al. [84] to estimate the impact of a single medicine while accounting for the feature space of currently available known drugs. Potential medications were collected from a cohort of patients, and each medication can be assigned to either the intervention group or the control group. Estimates of the two groups' confounders and disease progression were made, and the bias was eventually corrected by using LSTM in conjunction with the attention mechanism, leading to the discovery of a medication with therapeutic effects.

The research of medication repurposing heavily relies on machine learning techniques. Traditional machine learning techniques, which are mostly employed in the early stages of the process, include Logistic Regression, Random Forest, Support Vector machines, KNN, and RotatE, among others [83].

1.6 Challenges with AMR

1.6.1 Multiple messages, multiple audiences, and high expense

The goals of a communication process are audiences. They are the message's ultimate recipients, and based on how they interpret it, they may display overt behavior. A message is a concise, focused statement that often consists of a single major idea and is related to the communication goal [85]. The meticulous planning of communication methods, which are based on facts about audiences, communication theories, and audience behavior, is used to package messages [85,86]. It makes sense to undertake an audience analysis and segmentation when there are several audiences involved. For efficient communication, the various audience segments would need distinct message packaging and diverse media. This takes a lot of time and money.

1.6.2 AMR's novelty as a health communication issue

Few people paid attention to Sir Alexander Fleming's 1945 warning to the world about antibiotic resistance [87]. Since AMR is a novel occurrence of its kind, health communication specialists are uninitiated in this field. The difficulty with such first-time encounters is that you have to start almost from scratch, lacking experience and with less knowledge from small case studies that have been published. Before the issue worsens and becomes unmanageable, the globe must develop quickly in terms of AMR. Health communication specialists should provide guidance on communication and behavior change theories and concepts that should be tested, documented, and disseminated for immediate use.

KAP surveys must be conducted among a variety of audience groups in order to learn how people see AMR, identify dangerous behaviors, and package messages intended to modify such behaviors.

1.6.3 The part that poverty plays

The usage of antimicrobials is impacted by poverty, which can lead to dangerous behaviors such as sharing or storing for later use, which can lead to the development of resistance. In most parts of the world, sharing antibiotics between family members and neighbors is a regular practice [88,89]. Given that poverty has been linked to antibiotic usage, more research on this topic is warranted as it may reveal other characteristics of the target audience that have not yet been explored but may play a role in antimicrobial resistance.

1.6.4 Professional "abusers" who lack awareness

Health workers may bear some of the blame for AMR, even if they may not be entirely responsible for the breakout and spread of illnesses. Health professionals play a crucial role in Antimicrobial Use (AMU), and in order for them to effectively communicate AMR to their customers, they too need to become sensitized to the issue. Regarding AMU and AMR, they are reliable sources of information. When it comes to writing prescriptions for medication in the case of AMR, veterinarians and human doctors may succumb to client pressure. This audience needs to be converted first, since they offer both a challenge and a chance for AMR behavior change.

It is recommended that health professionals be trained in efficient clinician-patient communication so they may function as reliable sources and discuss the advantages and disadvantages of antibiotics for better clinical decision-making and patient satisfaction [90,91].

1.6.5 International versus domestic data

Statistics are useful tools in communication because they improve the attractiveness of messages to the intended audience, which aids in information packaging. Statistics demonstrate the need for an intervention. They make a compelling case for action. However, statistics need to be close to the intended consumers in terms of both geography and psychology. Local data are required for appropriate lobbying at the national and regional levels. Contextual and national statistics are necessary for policy makers to make informed decisions about how best to allocate resources for AMR. The target audience for statistics should be the country, not the world, as is the case with AMR at the moment.

1.7 AI Limitations

Though AI in healthcare has advanced to an accuracy level previously unthinkable, there are still many obstacles that make it challenging to integrate into care pathways. First, there is a certain mistrust among doctors regarding the use of AI-based systems due to the dearth of randomized clinical trials, particularly in the pediatric field, that show the dependability and/or increased effectiveness of AI systems compared to traditional systems in diagnosing infectious diseases or recommending suitable therapies. Health care professionals are underrepresented in the AI culture itself: many physicians, even pediatricians in our instance, are unaware of AI [92].

Another drawback of these systems is the potential for methodological bias because they frequently rely on databases, recommendations, and research from other nations that might not be entirely representative of all patients [93].

The requirement for a substantial quantity of data is another pertinent restriction. The amount of data required may rise with the complexity of the AI/ML architecture being utilized (deep neural networks, for example, need a lot of data). Furthermore, since NN and DNN are prone to overfitting data if the variability, such as variance, is restricted, the necessary data may not only need to be many but also of a high quality in terms of both data cleaning and variability. In some pediatric applications, it may be difficult, if not impossible, to gather a significant quantity of clean, validated, and varied data.

1.8 Conclusion

Since antimicrobial resistance is a complex issue, involvement from a range of stakeholders is needed. The creation of effective communication strategies should support the existence of robust national and global action plans for antimicrobial resistance. In contrast to other medical diseases, antimicrobial resistance (AMR) is a distinct phenomenon that necessitates the development of innovative mitigation techniques. Therefore, it is imperative that all AMR players use communication techniques that will result in behavior prediction and practical behavior modification tactics. In order to combat AMR, this will probably be a gradual process requiring significant and continuous participation from all players.

Novel drug candidates and drug discovery techniques are being made possible by methodologies that integrate systems and synthetic biology with machine learning models. Examples of such approaches include graph neural networks, sequence-to-function and sequence-to-structure frameworks, and generative models. In order to create AI models that are both biased-free and generalizable, large-scale datasets must be gathered and integrated. These datasets might comprise robust and representative genetic sequencing data, data from drug-target interactions and biomolecular interactions, and high-throughput therapeutic counter-screens and explorations of different chemical spaces for drug development.

Innate immunity requires antimicrobial peptides as a fundamental component. They could prove to be a good substitute for antibiotics. To accelerate the development of AMPs as medicines, it is imperative to understand the MOA that enables them to kill bacteria. Many different factors affect the selectivity and activity of these peptides. Net charge, hydrophobicity, secondary structure, and amphipathicity are a few examples of properties that are crucial for function and are so intertwined that modifying one usually affects the others.

References

[1] A. A. Rabaan *et al.,* "Application of Artificial Intelligence in Combating High Antimicrobial Resistance Rates," Antibiotics (Basel, Switzerland), vol. 11, no. 6, p. 784, Jun. 2022. doi:10.3390/antibiotics11060784

[2] A. Sakagianni *et al.,* "Using Machine Learning to Predict Antimicrobial Resistance-A Literature Review," Antibiotics (Basel, Switzerland), vol. 12, no. 3, p. 452, Feb. 2023. doi:10.3390/antibiotics12030452

[3] D. R. Giacobbe *et al.,* "Explainable and Interpretable Machine Learning for Antimicrobial Stewardship: Opportunities and Challenges," *Clinical Therapeutics,* Mar. 2024, doi:10.1016/j.clinthera.2024.02.010

[4] J. Lv, S. Deng, and L. Zhang, "A review of artificial intelligence applications for antimicrobial resistance," *Biosafety and Health,* vol. 3, no. 01, pp. 22–31, 2021.

[5] M.-H. Tran et al., "A New Hope in the Fight Against Antimicrobial Resistance with Artificial Intelligence," *Infection and Drug Resistance,* vol. 15, pp. 2685–2688, May 2022. doi:10.2147/IDR.S362356

[6] M. Chumbita et al., "Can artificial intelligence improve the management of pneumonia," *Journal of Clinical Medicine,* vol. 9, no. 1, p. 248, 2020.

[7] U. Schmidt-Erfurth *et al.,* "Machine learning to analyze the prognostic value of current imaging biomarkers in neovascular age-related macular degeneration," Ophthalmology Retina, vol. 2, no. 1, pp. 24–30, 2018.

[8] D. S. Sisodia, R. B. Pachori, and L. Garg, Eds., Handbook of Research on Advancements of Artificial Intelligence in Healthcare Engineering. IGI Global, 2020.
D. Ferrucci *et al.,* "Watson: beyond jeopardy!," Artificial Intelligence, vol. 199, pp. 93–105, 2013.

[9] M. Chumbita *et al.,* "Can artificial intelligence improve the management of pneumonia," *Journal of Clinical Medicine,* vol. 9, no. 1, p. 248, 2020.

[10] P. Kokol, J. Završnik, and H. B. Vošner, "Artificial intelligence and pediatrics: A synthetic mini review," arXiv preprint arXiv: 1802.06068, 2018.

[11] N. McCauley and M. Ala, "The use of expert systems in the healthcare industry," Information & Management, vol. 22, no. 4, pp. 227-235, 1992.

[12] "IBM's Watson AI Saves Woman's Life By Diagnosing Rare Form of Leukaemia," Available online: https: //www.huffingtonpost.co.uk/entry/ibms-

[13] "Computer Assisted Medical Diagnosis System for Inborn Errors of Metabolism," Available online: [14] https: //www.scopus.com/record/display.uri?eid=2-s2.0-0021404463&origin=inward&txGid=107d02549aa1348852833e63ffdd96a4

[15] "Watson the Pediatrician," Available online: https: //www. clinicalresearchnewsonline.com/news/2016/02/05/watson-the-pediatrician

[16[L. Deleger et al., "Developing and evaluating an automated appendicitis risk stratification algorithm for pediatric patients in the emergency department," *Journal of the American Medical Informatics Association,* vol. 20, no. e2, pp. e212-e220, 2013.

[17] S. Mani et al., "Medical decision support using machine learning for early detection of late-onset neonatal sepsis," Journal of the American Medical Informatics Association, vol. 21, no. 2, pp. 326-336, 2014.

[18] S. J. Patel, D. B. Chamberlain, and J. M. Chamberlain, "A machine learning approach to predicting need for hospitalization for pediatric asthma exacerbation at the time of emergency department triage," *Academic Emergency Medicine,* vol. 25, no. 12, pp. 1463-1470, 2018.

[19] A. Chaiboonchoe, S. Samarasinghe, and D. Kulasiri, "Using emergent clustering methods to analyse short time series gene expression data from childhood leukemia treated with glucocorticoids," 2009.

[20] B. M. Salazar et al., "Neuroblastoma, a paradigm for big data science in pediatric oncology," International Journal of Molecular Sciences, vol. 18, no. 1, p. 37, 2016.

[21] S. Ramgopal et al., "Seizure detection, seizure prediction, and closed-loop warning systems in epilepsy," *Epilepsy & Behavior,* vol. 37, pp. 291-307, 2014.

[22] A. Crippa et al., "Use of machine learning to identify children with autism and their motor abnormalities," *Journal of Autism and Developmental Disorders,* vol. 45, pp. 2146-2156, 2015.

[23] W. Liu, M. Li, and L. Yi, "Identifying children with autism spectrum disorder based on their face processing abnormality: A machine learning framework," *Autism Research,* vol. 9, no. 8, pp. 888-898, 2016.

[24] E. M. Kościuczuk et al., "Cathelicidins: family of antimicrobial peptides. A review," Molecular Biology Reports, vol. 39, no. 12, pp. 10957-70, 2012. doi:10.1007/s11033-012-1997-x

[25] P. Kumar et al., "Antimicrobial Peptides: Diversity, Mechanism of Action and Strategies to Improve the Activity and Biocompatibility *In Vivo*," *Biomolecules,* vol. 8, no. 1, p. 4, Jan. 2018. doi:10.3390/biom8010004

[26] J. Li et al., "Membrane Active Antimicrobial Peptides: Translating Mechanistic Insights to Design," *Frontiers in Neuroscience,* vol. 11, p. 73, Feb. 2017. doi:10.3389/fnins.2017.00073

[27] J. Lei *et al.,* "The antimicrobial peptides and their potential clinical applications," *American Journal of Translational Research,* vol. 11, no. 7, pp. 3919–3931, Jul. 2019.

[28] M. Pushpanathan *et al.,* "Antimicrobial peptides: versatile biological properties," *International Journal of Peptides,* vol. 2013, p. 675391, 2013. doi:10.1155/2013/675391

C. E. Jäkel *et al.,* "Efficacy of a proapoptotic peptide towards cancer cells," *In Vivo* (Athens, Greece), vol. 26, no. 3, pp. 419–426, 2012.

[29] Y. Huan, Y. Chuan, *et al.,* "Antimicrobial Peptides: Classification, Design, Application and Research Progress in Multiple Fields," Frontiers in microbiology, vol. 11, p. 582779, Oct. 2020. doi:10.3389/fmicb.2020.582779.

[30] H. Moravej *et al.*, "Antimicrobial Peptides: Features, Action, and Their Resistance Mechanisms in Bacteria," Microbial drug resistance (Larchmont, N.Y.), vol. 24, no. 6, pp. 747-767, 2018. doi:10.1089/mdr.2017.0392.

[31] L. J. Zhang and R. L. Gallo, "Antimicrobial eptides," Current Biology, vol. 26, no. 1, pp. R14–R19, 2016. doi:10.1016/j.cub.2015.11.017.

[32] R. Seyfi *et al.*, "Antimicrobial peptides (AMPs): roles, functions and mechanism of action," *International Journal of Peptide Research and Therapeutics,* vol. 26, pp. 1451–1463, 2020.

[33] L. J. Zhang and R. L. Gallo, "Antimicrobial peptides," *Current biology: CB,* vol. 26, no. 1, pp. R14–R19, 2016. doi:10.1016/j.cub.2015.11.017.

[34] Y. Huang *et al.*, "Alpha-helical cationic antimicrobial peptides: relationships of structure and function," *Protein & cell,* vol. 1, no. 2, pp. 143–152, 2010. doi:10.1007/s13238-010-0004-3.

[35] E. de Leeuw *et al.*, "Functional interaction of human neutrophil peptide-1 with the cell wall precursor lipid II," *FEBS letters,* vol. 584, no. 8, pp. 1543–1548, 2010. doi:10.1016/j.febslet.2010.03.004.

[36] A. Som, et al., "Synthetic mimics of antimicrobial peptides," *Biopolymers,* vol. 90, no. 2, pp. 83-93, 2008. doi:10.1002/bip.20970.

[37] G. Wang, "Antimicrobial peptides: discovery, design and novel therapeutic strategies," 2017.

[38] M. Zasloff, "Antimicrobial Peptides of Multicellular Organisms: My Perspective," Advances in experimental medicine and biology, vol. 1117, pp. 3–6, 2019. doi:10.1007/978-981-13-3588-4_1.

[39] K. Lohner, "Membrane-active Antimicrobial Peptides as Template Structures for Novel Antibiotic Agents," *Current Topics in Medicinal Chemistry,* vol. 17, no. 5, pp. 508-519, 2017.

[40] D. Münch and H.-G. Sahl, "Structural variations of the cell wall precursor lipid II in Gram-positive bacteria—Impact on binding and efficacy of antimicrobial peptides," *Biochimica et biophysica acta,* vol. 1848, no. 11 Pt B, pp. 3062-3071, 2015. doi:10.1016/j.bbamem.2015.04.014.

[41] F. Niyonsaba *et al.*, "A cathelicidin family of human antibacterial peptide LL-37 induces mast cell chemotaxis," *Immunology,* vol. 106, no. 1, pp. 20–26, 2002. doi:10.1046/j.1365-2567.2002.01398.x.

[42] R. Condé, *et al.*, "Natural antimicrobial peptides from eukaryotic organisms," Antimicrobial Agents, V. Bobbarala (ed), pp. 51–72, 2012.

[43] B. Mojsoska *et al.*, "Structure-activity relationship study of novel peptoids that mimic the structure of antimicrobial peptides," *Antimicrobial agents and chemotherapy,* vol. 59, no. 7, pp. 4112–4120, 2015. doi:10.1128/AAC.00237-15.

[44] H. Jenssen *et al.*, "Peptide antimicrobial agents," Clinical microbiology reviews, vol. 19, no. 3, pp. 491–511, 2006. doi:10.1128/CMR.00056-05.

[45] D. Juretić *et al.*, "Tools for Designing Amphipathic Helical Antimicrobial Peptides," Methods in molecular biology (Clifton, N.J.), vol. 1548, pp. 23-34, 2017. doi:10.1007/978-1-4939-6737-7_2.

[46] T.-H. Lee *et al.*, "Antimicrobial Peptide Structure and Mechanism of Action: A Focus on the Role of Membrane Structure," *Current topics in medicinal chemistry,* vol. 16, no. 1, pp. 25–39, 2016. doi:10.2174/1568026615666150703121700.

[47] J. Harder and J.-M. Schröder, Eds., Antimicrobial peptides: role in human health and disease. Springer, 2015.

[48] K. He *et al.*, "Neutron scattering in the plane of membranes: structure of alamethicin pores," *Biophysical journal,* vol. 70, no. 6, pp. 2659–2666, 1996.

[49] M. R. Yeaman and N. Y. Yount, "Mechanisms of antimicrobial peptide action and resistance," *Pharmacological reviews,* vol. 55, no. 1, pp. 27–55, 2003. doi:10.1124/pr.55.1.2.

[50] A. Matejuk *et al.*, "Peptide-based antifungal therapies against emerging infections," *Drugs of the Future,* vol. 35, no. 3, pp. 197, 2010.

[51] C. Stevenson, L. Marshall, and D. Morgan, Progress in Inflammation Research. Researchgate. Net, 2006, 216 p.

[52] T. Ganz, "Defensins: antimicrobial peptides of innate immunity," *Nature reviews. Immunology*, vol. 3, no. 9, pp. 710–720, 2003. doi:10.1038/nri1180.

[53] J. R. García *et al.*, "Identification of a novel, multifunctional beta-defensin (human beta-defensin 3) with specific antimicrobial activity. Its interaction with plasma membranes of Xenopus oocytes and the induction of macrophage chemoattraction," *Cell and tissue research,* vol. 306, no. 2, pp. 257–264, 2001. doi:10.1007/s004410100433.

[54] U. Fanelli *et al.*, "Improving the quality of hospital antibiotic use: Impact on multidrug-resistant bacterial infections in children," *Front. Pharmacol.,* vol. 11, p. 745, 2020. doi:10.3389/fphar.2020.00745.

[55] M. Komorowski, L. A. Celi, O. Badawi, A. C. Gordon, and A. A. Faisal, "The Artificial Intelligence Clinician learns optimal treatment strategies for sepsis in intensive care," *Nat. Med.,* vol. 24, pp. 1716–1720, 2018. doi:10.1038/s41591-018-0213-5.

[56] F. Lamping *et al.*, "Development and validation of a diagnostic model for early differentiation of sepsis and non-infectious SIRS in critically ill children—A data-driven approach using machine-learning algorithms," *BMC Pediatr.,* vol. 18, pp. 1-11, 2018. doi:10.1186/s12887-018-1082-2.

[57] H. Liang, et al., "Evaluation and accurate diagnoses of pediatric diseases using artificial intelligence," Nat. Med., vol. 25, pp. 433-438, 2019. doi:10.1038/s41591-018-0335-9.

[58] U. Fanelli *et al.*, "Role of Artificial Intelligence in Fighting Antimicrobial Resistance in Pediatrics," Antibiotics (Basel, Switzerland), vol. 9, no. 11, p. 767, Nov. 2020. doi:10.3390/antibiotics9110767.

[59] M. Beaudoin, F. Kabanza, V. Nault, and L. Valiquette, "Evaluation of a machine learning capability for a clinical decision support system to enhance antimicrobial stewardship programs," *Artif. Intell. Med.*, vol. 68, pp. 29–36, 2016. doi:10.1016/j.artmed.2016.02.001.

[60] U. Fanelli, V. Chiné, M. Pappalardo, P. Gismondi, and S. Esposito, "Improving the quality of hospital antibiotic use: Impact on multidrug-resistant bacterial infections in children," *Front. Pharmacol.*, vol. 11, p. 745, 2020. doi:10.3389/fphar.2020.00745.

[61] M. Beaudoin, F. Kabanza, V. Nault, and L. Valiquette, "An antimicrobial prescription surveillance system that learns from experience," *AI Mag.*, vol. 35, pp. 15–25, 2014. doi:10.1609/aimag.v35i1.2500.

[62] M. Oonsivilai *et al.*, "Using machine learning to guide targeted and locally-tailored empiric antibiotic prescribing in a children's hospital in Cambodia," *Wellcome Open Res.*, vol. 3, p. 131, 2018. doi:10.12688/wellcomeopenres.14847.1.

[63] P. M. Kilbridge *et al.*, "Computerized surveillance for adverse drug events in a pediatric hospital," *J. Am. Med. Informatics Assoc.*, vol. 16, pp. 607–612, 2009. doi:10.1197/jamia.M3167.
[64] S. A. Baltzer and M. H. Brown, "Antimicrobial peptides—Promising alternatives to conventional antibiotics," *J. Mol. Microbiol. Biotechnol.*, vol. 20, pp. 228–235, 2011. doi:10.1159/000331009.
[65] J. Kim *et al.*, "VAMPr: VAriant Mapping and Prediction of antibiotic resistance via explainable features and machine learning," *PLoS Comput. Biol.*, vol. 16, p. e1007511, 2020. doi:10.1371/journal.pcbi.1007511.
[66] CRyPTIC Consortium and the 100,000 Genomes Project, C. Allix-Béguec et al., "Prediction of susceptibility to first-line tuberculosis drugs by DNA sequencing," *N. Engl. J. Med.*, vol. 379, pp. 1403–1415, 2018.
[67] S. Yamaguchi, M. Kaneko, and M. Narukawa, "Approval success rates of drug candidates based on target, action, modality, application, and their combinations," *Clin. Transl. Sci.*, vol. 14, pp. 1113–1122, 2021.
[68] M. Schlander *et al.*, "How Much Does It Cost to Research and Develop a New Drug? A Systematic Review and Assessment," *Pharmacoeconomics*, vol. 39, pp. 1243–1269, 2021.
[69] J. Söding, A. Biegert, and A. N. Lupas, "The HHpred interactive server for protein homology detection and structure prediction," *Nucleic Acids Res.*, vol. 33, pp. W244–W248, 2005.
[70] L. Zimmermann *et al.*, "A Completely Reimplemented MPI Bioinformatics Toolkit with a New HHpred Server at its Core," *J. Mol. Biol.*, vol. 430, pp. 2237–2243, 2018.
[71] D. S. Marks *et al.*, "Protein 3D structure computed from evolutionary sequence variation," *PLoS ONE*, vol. 6, no. 9, p. e28766, 2011.
[72] S. Wang *et al.*, "Accurate De Novo Prediction of Protein Contact Map by Ultra-Deep Learning Model," *PLoS Comput. Biol.*, vol. 13, no. 1, p. e1005324, 2017.
[73] Z. Du *et al.*, "The trRosetta server for fast and accurate protein structure prediction," *Nat. Protoc.*, vol. 16, pp. 5634–5651, 2021.
[74] J. Yang and Y. Zhang, "Protein Structure and Function Prediction Using I-TASSER," *Curr. Protoc. Bioinform.*, vol. 52, pp. 5.8.1–5.8.15, 2015.
[75] J. Moult *et al.*, "Critical assessment of methods of protein structure prediction (CASP)-Round XII," *Proteins*, vol. 86, pp. 7–15, 2018.
[76] R. A. Laskowski *et al.*, "AQUA and PROCHECK-NMR: Programs for checking the quality of protein structures solved by NMR," *J. Biomol. NMR*, vol. 8, pp. 477–486, 1996.
[77] F. Melo, D. Devos, E. Depiereux, and E. Feytmans, "ANOLEA: A www server to assess protein structures," in *Proc. Int. Conf. Intell. Syst. Mol. Biol.*, 1997, pp. 187–190.
[78] F. Yang *et al.*, "Machine Learning Applications in Drug Repurposing," *Interdiscip. Sci. Comput. Life Sci.*, vol. 14, no. 1, pp. 15–21, 2022.
[79] Z. Wu, Y. Wang, and L. Chen, "Network-based drug repositioning," *Mol. BioSyst.*, vol. 9, no. 6, pp. 1268–1281, 2013.
[80] E. Kim, A. Choi, and H. Nam, "Drug repositioning of herbal compounds via a machine-learning approach," *BMC Bioinformatics*, vol. 20, pp. 33–43, 2019.

[81] R. Liu, L. Wei, and P. Zhang, "A deep learning framework for drug repurposing via emulating clinical trials on real-world patient data," *Nat. Mach. Intell.*, vol. 3, no. 1, pp. 68–75, 2021.
[82] S. D. Ferguson, *Communication Planning: An Integrated Approach.* Sage, Thousand Oaks, CA, 1999.
[83] A. H. Eagly and S. Chaiken, *The psychology of attitudes.* New York: Harcourt, Brace, & Janovich, 1993.
[84] A. Fleming, "Penicillin's finder assays its future," *New York Times*, 1945, Jun. 26. [Online]. Available: https: //www.nytimes.com/1945/06/26/archives/penicillins-finder-assays-its-future-sir-alexander-fleming-says.html. [Accessed: 01-Mar-2020].
[85] M. B. Planta, "The role of poverty in antimicrobial resistance," *J. Am. Board Fam. Med.*, vol. 20, no. 6, pp. 533–539, 2007.
[86] C. A. Hart and S. Kariuki, "Antimicrobial resistance in developing countries," *BMJ*, vol. 317, pp. 647–650, 1998.
[87] P. Coxeter, C. B. Del Mar, and L. McGregor, "Interventions to facilitate shared decision making to address antibiotic use for acute respiratory infections in primary care," *Cochrane Database Syst. Rev.*, vol. 11, p. CD010907, 2015.
[88] D. Nutbeam and I. Kickbusch, "Advancing health literacy: a global challenge for the 21st century," *Health Promot. Int.*, vol. 15, pp. 183–184, 2000.
[89] P. Kokol, J. Završnik, and H. B. Vošner, "Artificial intelligence and pediatrics: A synthetic mini review," *arXiv preprint arXiv: 1802.06068*, 2018.
[90] "Bias Test to Prevent Algorithms Discriminating Unfairly," *New Scientist.* [Online]. Available: https: //www.newscientist.com/article/mg23431195-300-bias-test-to-prevent-algorithms-discriminating-unfairly/. [Accessed: 8-May-2020].
[91] "Google DeepMind 1.6 m Patient Record Deal 'Inappropriate'," *The Guardian.* [Online]. Available: https: //www.theguardian.com/technology/2017/may/16/google-deepmind-16m-patient-record-deal-inappropriate-data-guardian-royal-free. [Accessed: 8-May-2020].
[92] F. Cabitza, C. Alderighi, R. Rasoini, and G. F. Gensini, "Handle with care': About the potential unintended consequences of oracular artificial intelligence systems in medicine," *Recenti Prog. Med.*, vol. 108, pp. 397–401, 2017.
[93] "Facebook Shuts Down Robots After They Invent Their Own Language," *The Telegraph.* [Online]. Available: https: //www.telegraph.co.uk/technology/2017/08/01/facebook-shuts-robots-invent-language/. [Accessed: 8-May-2020].

2

Artificial Intelligence in Disease Surveillance and Monitoring Resistance Microbes

Gajala Deethamvali Ghouse Peer,[1] *Arpana Vibhuti,*[2] *Gajala Ishrathtaj*[3] and *Ramendra Pati Pandey*[1,*]

The rapid outbreak monitoring to prevent widespread health crises have become necessitates for the emergence of numerous infectious diseases especially Anti Microbial Resistance. Effective disease control relies on surveillance methods tailored to clinical approaches and epidemiology, considering evolving pathogens, at-risk populations, and ongoing mortality and morbidity data. Drug-resistant microorganisms have surged, compromising treatment efficacy and threatening global healthcare systems. Artificial Intelligence (AI) offers transformative potential in healthcare and microbiology by improving diagnostic accuracy, expediting drug discovery, and combating antimicrobial resistance. AI's rapid analysis of genetic sequences enhances pathogen identification, while its predictive capabilities streamline drug development, thus supporting global health efforts. This chapter explores AI's role in monitoring resistant microbes and disease surveillance.

[1] School of Health Science and Technology (SOHST), UPES, Dehradun. Email: gajala.127662@stu.upes.ac.in
[2] Department of Biotechnology, SRM University Delhi-NCR, Sonepat.
[3] Department of Chemistry, SRM University, Delhi-NCR, Sonepat. Email: gajalaishrathtaj@gmail.com
* Corresponding author: ramendra.pandey@gmail.com

2.1 Introduction

The emergence of numerous infectious diseases in the twenty-first century has made it essential to quickly monitor outbreaks before they cost lives and spread throughout the community. To control any disease or disease outbreak, surveillance is important to keep track of different approaches that can create impact in controlling the diseases based on their clinical approach and disease epidemiology. A successful effort to control infectious diseases must take into account the evolving nature of disease agents, populations at risk, and ongoing observation of mortality and morbidity [1]. Resistance among microorganisms against drugs has been increased in the last decade and causing a rise in the lack of effectiveness of drugs in developed and developing nations. The Rapidly evolving nature of these resistant microbes has become difficult to treat, which is threatening to modern healthcare, medical systems, raising the risk of infection, serious medical problems, and death. This drug resistance makes antimicrobial drugs, including antibiotics, ineffective and makes treating infections challenging or impossible [2].

Artificial Intelligence (AI) holds tremendous promise for transforming healthcare and microbiology by enhancing diagnostic accuracy, accelerating drug discovery, and combating antimicrobial resistance. It has advanced quickly in the last two decades, transforming a wide range of sectors, including microbiology and healthcare. AI has great potential to advance diagnosis, treatment, medication discovery, and preventative medicine in several domains. One key area is the identification and classification of microorganisms. Traditional methods for microbial identification, such as culturing and biochemical assays, are time-consuming and labor-intensive. AI algorithms, however, can analyze genetic sequences from microbial samples rapidly and accurately, enabling swift identification of pathogens and facilitating targeted interventions in disease outbreaks [3].

Furthermore, AI can enhance drug discovery processes by predicting the efficacy and safety of potential drug candidates. Pharmaceutical companies are increasingly leveraging AI to screen vast libraries of compounds, predict their interactions with biological targets, and design novel molecules with desired pharmacological properties. This approach expedites the drug development pipeline and reduces the cost of bringing new medicines to market, ultimately benefiting patients worldwide [4]. This chapter mainly focuses on the differential role of AI in monitoring the resistance microbes and surveillance of different diseases that are originated by these microbes.

2.2 Understanding Disease Surveillance

Disease surveillance is an example of information-based work that entails obtaining, evaluating, and interpreting enormous volumes of data from numerous sources. Public health surveillance provides the actual and scientific database required for informed choices and appropriate public health action. The primary objective of surveillance is to acquire information to guide interventions. The aims of public health and the actions necessary to conduct successful interventions determine the design and implementation of surveillance systems [5]. Disease surveillance refers to the systematic collection, analysis, interpretation, and dissemination of data on diseases and health-related events [6]. It is essential to public health because it makes early disease outbreak detection and monitoring possible, enables quick reaction and control measures, and aids in the identification of patterns and trends in disease occurrence. Disease surveillance minimizes the spread of infections and lowers the overall burden on communities by tracking and identifying diseases early on, assisting public health officials in intervening and putting preventive measures into action. Additionally, disease surveillance offers useful data for policy-making, resource allocation, and public health planning [6,7,8]. With the use of this data, focused public health initiatives or vaccination campaigns can be developed and implemented to effectively address certain disease risks [6].

To improve public health, disease surveillance makes it possible to detect and track diseases early on, respond quickly to them, take appropriate action, spot patterns and trends, and plan and implement public health interventions. Furthermore, disease surveillance makes it possible to identify newly discovered illnesses, which is essential for averting pandemics and outbreaks. In conclusion, disease surveillance is an essential public health tool that supports early outbreak detection, disease identification and monitoring, prompt response measures, and the development of public health plans and actions. In public health, disease surveillance is essential because it makes early disease detection, monitoring, and control possible [9].

It is impossible to exaggerate the significance of disease surveillance in public health. It performs several vital roles that are necessary for protecting and improving population health, including **Early Detection**

of Outbreaks, where Disease Surveillance makes it possible to identify outbreaks and epidemics promptly, enabling public health authorities to take swift action to stop the spread of the disease. Prompt detection of epidemics is essential to reduce morbidity, mortality, and disturbance to society [10].

Monitoring disease trends is an additional essential function, which involves using surveillance data to offer light on historical trends and epidemiological patterns in diseases. Public health experts can spot new health risks, evaluate the success of interventions, and deploy resources wisely by keeping an eye on changes in disease incidence, prevalence, and distribution [11].

Assessing Public Health Impact is the third crucial role. Disease surveillance makes it possible to assess the results of public health policies and initiatives. Public health authorities can evaluate the success of control measures, pinpoint areas for improvement, and allocate resources as efficiently as possible by comparing pre- and post-intervention data [12].

The surveillance data that forms the basis of evidence-based decision-making in public health is known as **Supporting Evidence-Based Decision-Making**. Surveillance data is used by policymakers, healthcare professionals, and other stakeholders to prioritize health needs, distribute resources, create intervention plans, and create public health policies.When it comes to **Informing Public Health Response** where Disease surveillance provides real-time information that informs the development and implementation of targeted public health responses. By identifying populations at increased risk, geographic hotspots, and modes of transmission, public health authorities can tailor interventions to specific needs and mitigate health disparities [5].

Detecting Changes in Disease Patterns makes Surveillance systems essential for detecting changes in disease patterns, including the emergence of new pathogens, the re-emergence of previously controlled diseases, and shifts in antimicrobial resistance patterns. Early detection of such changes is critical for initiating timely responses and preventing widespread outbreaks [13].

A framework for reviewing literature on Public Health Surveillance (PHS) is shown in Fig. 2. The three main parts of the PHS system are as follows: purpose, performance attributes, and data. Any surveillance system is built on data, which comes from a variety of sources including social media, the internet, Electronic Health Records (EHR), and public observational cohorts. The extent, resolution, and nature of

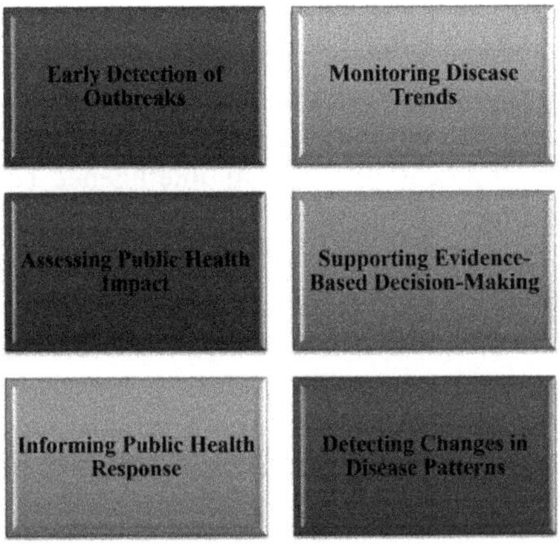

Figure 1: Vital roles Disease surveillances.

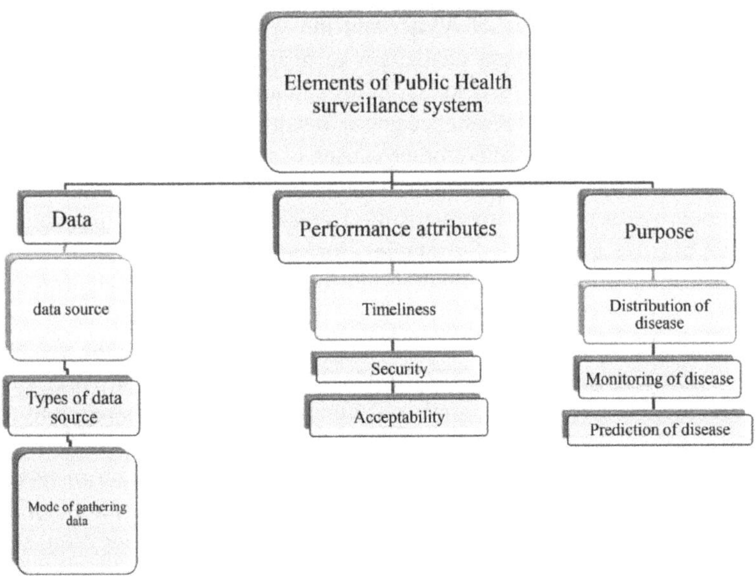

Figure 2: A framework for reviewing literature on public health surveillance [14].

the data—manually gathered or event-triggered—are all taken into account. Consent from the population, data security, and timeliness are performance factors that are evaluated. The technology may be used for disease dissemination, monitoring, or data-driven spread prediction in the future. Systems can integrate different types, objectives, and data sources to maximize performance characteristics.

The need for advanced technologies like AI to enhance surveillance capabilities would rather help to improve the disease surveillance system. The development of artificial intelligence (AI) technologies predicts an important development in the control and prevention of infectious diseases [15]. Among other areas of epidemic control, this advancement involves improved early outbreak identification, contact tracking, infection diagnosis, medication discovery, and assistance with drug formulation..

2.3 Overview of Antimicrobial Resistance (AMR)

Since antibiotics are misused and overused in many different contexts, leading to the evolution of resistant bacteria, A ntimicrobial R esistance (AMR) is a serious worldwide health concern. The discovery of penicillin is the source of AMR, and the spread of bacteria resistant to drugs poses significant challenges to healthcare systems across the globe. Antibiotic overuse in agriculture, human health, and animals promotes the spread of resistance genes, resulting in a "Silent Pandemic" that is predicted to overtake other causes of death by 2050 [16]. AMR affects both humans and animals, and treating infections is made more difficult by resistant microorganisms. To resist the effects of antibiotics, microbes use a variety of strategies, including enzyme alteration and the creation of biofilms. Though issues like data quality and algorithmic biases still exist, artificial intelligence is being investigated as a way to improve diagnosis and treatment approaches in the fight against antimicrobial resistance (AMR). A One Health strategy that takes into account environmental, animal, and human components is required to effectively treat AMR. This entails supporting stewardship initiatives, strengthening surveillance systems, and funding the creation of novel antibiotic treatments. In order to combat AMR and ensure that antibiotics remain effective for future generations, public awareness, education, and international cooperation are essential [17].

AMR has become recognized as a complicated issue that impacts both human and animal health. The misuse and improper application of antibiotics in an assortment of industries, such as veterinary medicine, agriculture, and hospital settings, has sped up the emergence of microbial resistance. The overuse of antibiotics has led to the emergence of "superbugs," or antibiotic-resistant bacteria, which pose significant challenges to the efficacy of treatment and may cause life-threatening infections. Furthermore, as the development of resistance outpaces the discovery of effective therapies, the issue is made worse by the inadequate progress made in the development of new antimicrobial drugs [18].

The 21st century has seen the emergence of antimicrobial resistance (AMR) as a major danger to human health, with formerly preventable diseases growing more resistant resulting in substantial therapeutic challenges. Growing dependence on second and third-line therapies which frequently have higher costs, higher toxicity, and longer treatment durations has resulted from the exhaustion of first-line antimicrobial medicines. As a result of protracted hospital stays brought on by prolonged diseases, resources for both individuals and the healthcare system are depleted [19]. Furthermore, longer recovery times reduce economic production by requiring workers to take longer breaks from work. Increased laboratory testing, more frequent outpatient clinic visits, and stricter isolation protocols are also necessary in response to AMR-related infections. Over a million deaths are directly linked to AMR pathogens every year [20].

Because infections become more difficult to treat, the availability of strong antibiotics puts common medical treatments at risk, including organ transplants, chemotherapy, surgeries, and neonatal care. Multidrug-resistant pathogens, also referred to as "**ESKAPE**" include resistant strains of *Enterococcus faecium, Staphylococcus aureus, Klebsiella pneumoniae, Acinetobacter baumannii, Pseudomonas aeruginosa,* and *Enterobacter* spp. [17]. These bacteria pose a serious threat in hospital settings today.

Enhancing the effectiveness and efficiency of public health policies and decision-making processes is largely dependent on surveillance systems. As surveillance systems help healthcare professionals, caregivers, and policymakers make informed decisions to improve healthcare delivery based on real-time data, this largely entails increasing population health. The three main roles of public health services are typically divided into, policy development, assessment and assurance [21]. Even if a surveillance system might support one of these goals, it is crucial that every surveillance system clearly identify its unique purpose. Furthermore, belief surveillance can be used to gauge users' confidence in the public health system [22], which advances the public health goal of assurance.

2.4 Role of Artificial Intelligence in Disease Surveillance

Artificial intelligence (AI) techniques are widely used in the prediction of trends, public health response modeling, early infectious disease outbreak detection and alerting, and trend forecasting. Limited data, a lack of positive training examples, difficulties setting benchmarks and measuring control measures, intricate relationships between spatial and temporal factors, and risk assessments through contact and social networks are some of the unique technical challenges these crucial public health surveillance tasks present. While statistical methods are still the mainstay of traditional public health surveillance, recent years have seen a notable increase in AI-based techniques, such as deep learning models, which serve to supplement statistical methods.

2.4.1 Application of AI in analyzing large datasets for disease detection and monitoring

AI introduces novel opportunities in public health surveillance by leveraging diverse data sources, including those not initially intended for epidemiological inquiry. For example, the proliferation of Internet and Internet of Things technologies (IOT) enables widespread social and device sensing capabilities, unlocking significant surveillance potential. Utilizing open data beyond traditional surveillance systems can augment surveillance capabilities effectively. Additionally, AI enhances traditional data analytics tools, predominantly rooted in statistics, to accommodate diverse data types, including unstructured text, images, and videos, alongside structured data. Handling unstructured data necessitates AI techniques like natural language processing and image processing, often employing deep learning for automated feature extraction and labeling [23]. Analytically, public health surveillance aims to timely assess epidemic risk, detect spatial and temporal abnormalities, and provide early warnings and trend predictions. Machine learning-based AI methods have long been employed to identify patterns, anomalies, and trends in public health data streams, which often exhibit pronounced temporal and spatial characteristics and require integration with external socioeconomic and environmental data. In contrast, traditional surveillance methods, heavily reliant on statistical approaches, struggle with the increasing complexity of data, offering limited insights at a macro level. AI, particularly multiagent systems, provides modeling frameworks capable of simulating intricate infectious disease transmission dynamics, allowing for scenario exploration and quantitative assessment of intervention effects [23].

By assisting in the prediction of infectious diseases' spatial and temporal progression and dispersion, machine-learning algorithms contribute to the management of infectious diseases [24]. Large, complex datasets are no problem for these algorithms to process; they can reveal patterns and trends that a human observer might miss. Their capacity to foresee infectious diseases, which usually entail a variety of elements such as human behaviors, environmental conditions, and population demographics, makes them successful. With encouraging results, a number of researchers have used machine-learning techniques to forecast infectious diseases. However, the availability of comprehensive, high-quality data is a major obstacle to using machine learning for disease forecasting. Systems for monitoring infectious diseases usually collect information on a range of topics, including the demographics of affected people, the locations of outbreaks, and the reported case counts. However, these data frequently have biases, incompleteness, or noise, which affects how well machine-learning models work. Additionally, infectious diseases have protracted incubation periods, which causes differences between data from previous outbreaks and the state of affairs today [25].

There is potential for impact on public health, epidemiology, and clinical practice from the investigation of the integration of machine learning into infectious disease diagnosis. However, there are significant obstacles and restrictions associated with its use, such as the requirement for high-quality data, the complexity of the underlying phenomena, and the vulnerability to bias and overfitting [26]. It is difficult to obtain high-quality data because of problems such as incomplete records, missing values, and different data formats. Furthermore, the construction of accurate models is complicated by the complex dynamics involved in the spread of infectious diseases. Prediction accuracy is further threatened by the possibility of bias and overfitting in machine-learning frameworks [26].

2.4.2 AI-powered surveillance systems

The World Health Organization's (WHO) Global Action Plan (GAP), the new European One Health Action Plan against AMR, and the Central Asian and Eastern European Surveillance of Antimicrobial Resistance (CAESAR) network are just a few of the international initiatives that have been put into place to address the AMR crisis [27]. Antimicrobial

Usage (AMU) and Antimicrobial Resistance (AMR) monitoring systems in humans and animals are essential for assessing and controlling global trends in antibiotic use and bacterial susceptibility patterns in various populations. In a One Health framework, zoonotic and indicator microbes are essential.Antimicrobial Usage (AMU) and Antimicrobial Resistance (AMR) surveillance and monitoring systems in humans and animals are essential for assessing and controlling global trends in antibiotic use and bacterial susceptibility patterns in various populations. Diagnostic and zoonotic microbes are especially important in a One Health approach. The One Health idea was supported and advanced by a strategic framework for lowering the risks of infectious diseases at the animal-human ecological interface that was released in 2008. Since then, the One Health Approach has been adopted and put into practice by several national and international organizations [28,29].

2.5 AI Technologies for Microbial Monitoring and AMR Detection

With the quick emergence of strains that are resistant to several drugs and the protracted process of creating new antibiotics, bacterial antimicrobial resistance (AMR) represents a serious danger to national and international health. As a result, **Machine learning**, a subset of artificial intelligence, has gained attention as a promising way to stop the spread of antimicrobial resistance. AI can accurately forecast and identify AMR in bacteria. The discovery of new antimicrobials can also be sped up by combining laboratory testing and machine learning methods. To detect antibiotic resistance, an abundance of machine learning algorithms have been created and put through extensive testing. These algorithms have shown excellent accuracy and surpass conventional techniques that rely on sequence comparison databases.

2.5.1 AMR identification by AI

Before artificial intelligence (AI) was used to identify AMR determinants, antimicrobial susceptibility testing (AST) or bioinformatics comparison of sequences against databases were the primary methods of identification. Because their sequences differed significantly from known AMR sequences, these methods frequently resulted in a large proportion of actual AMR determinants being incorrectly categorized as non-AMR (false negatives). In addition, it takes a while to cultivate bacteria in AST, and some bacteria are difficult to grow *in-vitro* [30]. A new, quick, and effective way to reliably discover known and novel AMR determinants is provided by the *In-silico* methodology that makes use of Machine learning and model algorithms trained on labeled data. Afterward, the algorithm could analyze test data and autonomously label them.

2.5.2 Algorithms used in the identification of antimicrobial résistance gene from DNA sequence

There are a number of algorithms for identifying antibiotic resistance genes (ARGs), such as the Mykrobe predictor [31], which comes in the form of easily navigable software that can be downloaded right away. This method quickly determines whether bacteria are susceptible to a certain antibiotic or not by analyzing full genome sequences. Its application to only Mycobacterium TB and Staphylococcus aureus is a limitation, though. On the other hand, DeepARG [32] uses both short-sequence reads and full-length ORFs to predict ARGs. DeepARG may miss less common ARGs, such as those granting resistance to triclosan [33] and sulfonamide [34], even if it shows great specificity and sensitivity in finding ARGs. Though it would benefit from more training data containing lesser-known ARGs, DeepARG nevertheless exhibits potential as an online tool for discovering new ARGs [32]. Three other approaches predict AMR by analyzing short oligonucleotides (*kmers*) without prior knowledge of bacterial antimicrobial resistance genes, in addition to the method of using whole-genome sequences and associated antimicrobial resistance (AMR) metadata to forecast AMR phenotypes [35]. *Kmers* with a typical length of 10 to 40 bases are employed by these three widely used algorithms: XGBoost-based machine learning model [35], Kover [36] and RASTtk [37].

Recently, antimicrobial resistance (AMR) genes in both Gram-positive and Gram-negative bacteria have been identified using the Support Vector Machine (SVM) algorithm. This algorithm has shown the ability to predict AMR genes in a variety of bacteria, including *E. coli* [39,40], *M. tuberculosis* [41], *S. aureus, and P. aeruginosa* [40]. It uses pan-genome datasets and reference genomes as training data [42]. Predicting genetic interactions among various allele combinations by evaluating their impact on resistance, SVM searches pan-genomes for AMR-determining alleles, much like earlier algorithms that required antimicrobial resistance gene datasets. The number of antimicrobial resistance genes present in the microbe has been found to significantly

Table 1: Algorithms used to Identify AMR genes from DNA sequence [38].

Algorithms	**Usability of the algorithm**	**Functional**
DeepARG	The application runs effortlessly on personal PCs and produces quick results. It is also very user-friendly and simple to use.	Functions for all microorganisms; results for underrepresented ARGs are less accurate.
Mykrobe predictor	The application runs effortlessly on personal PCs and produces quick results (in around three minutes). It is also very user-friendly and simple to use.	Only functions with *S. aureus* and *M. tuberculosis*; results for underrepresented ARG are less accurate.
RASTtk	It is also very user-friendly and simple to use.	Only effective with specific pathogen-antimicrobial combinations.
Kover	Although there is a thorough tutorial explaining how to install and use the program, users must possess some IT knowledge. Users must also train the model. The tool is not user-friendly and difficult to manage.	Functions with any type of microorganism provided that users provide their own training datasets. Not as accurate of a result for underrepresented ARGs
Support Vector Machine	There is no online tool available.	Effective only against specific infections The research created pen-genomes. Not as accurate of a result for underrepresented ARGs.
XGBoost-based machine learning model	The tool is not user-friendly and difficult to manage.	Functions with any type of microorganism provided that users provide their own training datasets.
PCM	The tool is not user-friendly and difficult to manage.	Most protein sequences are compatible. Although the results are valid, the imbalanced dataset may have produced a skewed outcome because there are less negative references than positive references.
VAMPr model	It is also very user-friendly and simple to use.	Works only with nine species of bacteria, however you can train your model. Not as accurate of a result for underrepresented antibiotic-pathogen combinations.

Table 1: contd...

Table 1: contd...

Algorithms	Usability of the algorithm	Functional
GTDWFE	There is no online tool available.	Only functions with the protein sequences *bla, dfr*, and *aac*. The result's accuracy is fairly high (maybe as a result of using just well-known AMR proteins as datasets).

affect the algorithm's accuracy, exhibiting a wide range of sensitivity and specificity. For example, using larger antimicrobial resistance gene datasets has led to higher accuracy. The most recent version of the VAMPr algorithm, which is similar to XGBoost, necessitates developing unique models for different pathogen-antimicrobial pairings. It provides thorough insights by connecting ARGs with their protein structures through the incorporation of gene orthologs. It takes time to create models for every species-antibiotic combination [43]. All these algorithms are referred to in Table 1.

2.6 Artificial Intelligence in Combating High Antimicrobial Resistance Rates

To reduce the spread of resistant bacterial strains, organizations such as the World Health Organization (WHO) stress the need of selecting the right antibiotic and providing treatment for the least amount of time. So establishing an institutional antibiotic stewardship program is essential to combating the rising rates of antimicrobial resistance (AMR). This program would generate the antibiograms, control antibiotic distribution, and monitor the correct administration of antibiotics. These instruments may also be extremely helpful in emergency medical scenarios where prompt treatment is required without waiting for the findings of a bacterial culture. AI has a wide range of potential uses in healthcare, including cost-saving measures, improving diagnostic and therapeutic precision, and accelerating the development of novel antimicrobial medications. The majority of AI-based AMR solutions that have been put forward are meant to be helpful tools that help doctors work more efficiently, not to take the place of their recommendations or assessments [44]. Artificial Intelligence has the potential to be a game-changer in the fight against antibiotic resistance in the field of infectious diseases. Furthermore, to guarantee timely and efficient treatment, information from regional antibiotic stewardship programs is crucial for choosing antibiotic medication.

Combating antibiotic resistance requires prompt detection of pathogenic infections, the ability to differentiate between infectious and noninfectious symptoms and the administration of appropriate treatment. AI is essential for resolving this worldwide problem. For predicting antimicrobial resistance (AMR), using antibiograms and customized machine learning (ML) as discussed, based models may be crucial in detecting high-risk infectious organisms and monitoring trends in susceptibility [45], this is diagrammatically represented in Fig. 3. A similar study by Yelin et al. examined a dataset spanning ten

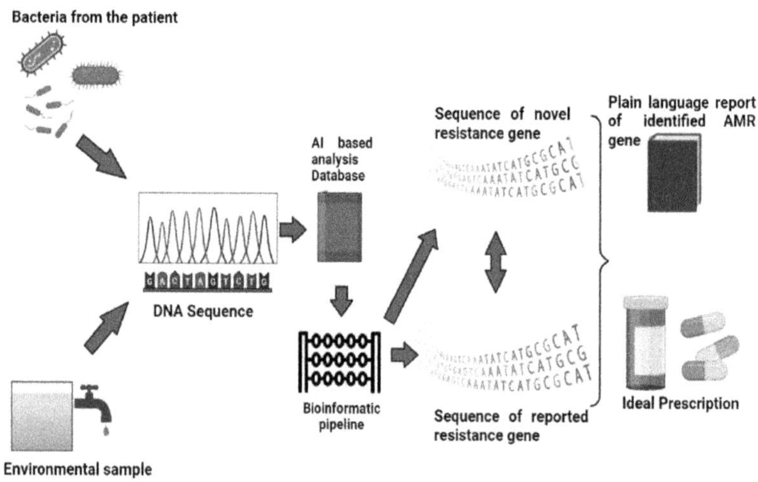

Figure 3: Gene-sequencing model with deep antibiotic resistance [44].

years that included more than 0.7 million. Their research showed a significant relationship between antibiotic use, past urine culture history, and demographic characteristics and AMR. They then created an AMR prediction model based on machine learning to determine the resistance patterns of common UTI bacteria [46].

2.7 Conclusion

In conclusion, quick surveillance is essential in the fight against infectious diseases in order to prevent outbreaks and lessen their effects. In this quest, artificial intelligence (AI) emerges as a crucial tool with unparalleled potential in microbiology and healthcare. Artificial intelligence algorithms facilitate prompt pathogen detection and targeted interventions during outbreaks by examining genetic sequences from microbial samples quickly. Additionally, AI speeds up drug discovery by forecasting the safety and efficacy of viable candidates, cutting down on the time and expense needed to introduce novel medications to the market. By analyzing large datasets, assisting with early outbreak identification, and offering real-time insights into disease trends and patterns, artificial intelligence (AI) enhances conventional methods in the field of disease monitoring.

It is impossible to overestimate the importance of disease surveillance in public health since it makes evidence-based decisions possible, tracks disease trends, and allows for the early discovery of outbreaks. By utilizing a variety of data sources and predictive modeling approaches, artificial intelligence (AI) is integrated into surveillance systems to improve their performance. Although issues such as algorithmic biases and poor data quality still exist, AI has the potential to completely transform disease surveillance and response operations. AI is essential in the fight against antimicrobial resistance (AMR) since it can anticipate and identify resistance patterns, speed up medication discovery, and help with antibiotic stewardship initiatives.Healthcare practitioners can minimize the spread of resistant diseases, optimize antibiotic usage, and make well-informed treatment decisions by utilizing AI's capabilities. The promise of AI must be weighed against ethical issues and continued efforts to solve algorithmic and data quality issues. All things considered, AI is a revolutionary force in disease surveillance and AMR management, providing creative answers to challenging problems in public health. AI-driven strategies can ease the way for a more robust and healthy future by fostering cooperation between academics, legislators, and healthcare professionals. Demonstrating how AI can transform antimicrobial resistance and disease surveillance. promoting more study and cooperation in this area to successfully handle global health issues.

References

[1] J. Murray and A. L. Cohen, "Infectious Disease Surveillance," *Int. Encycl. Public Heal.*, p. 222, Oct. 2017, doi:10.1016/B978-0-12-803678-5.00517-8.

[2] S. Ajulo and B. Awosile, "Global antimicrobial resistance and use surveillance system (GLASS 2022): Investigating the relationship between antimicrobial resistance and antimicrobial consumption data across the participating countries," *PLoS One*, vol. 19, no. 2, Feb. 2024, doi:10.1371/JOURNAL.PONE.0297921.

[3] C. Kooli and H. Al Muftah, "Artificial intelligence in healthcare: a comprehensive review of its ethical concerns," *Technol. Sustain.*, vol. 1, no. 2, pp. 121–131, Sep. 2022, doi:10.1108/TECHS-12-2021-0029/FULL/XML.

[4] T. Davenport and R. Kalakota, "The potential for artificial intelligence in healthcare," *Futur. Healthc. J.*, vol. 6, no. 2, p. 94, Jun. 2019, doi:10.7861/FUTUREHOSP.6-2-94.

[5] P. Nsubuga *et al.*, "Public Health Surveillance: A Tool for Targeting and Monitoring Interventions," *Dis. Control Priorities Dev. Ctries.*, 2006, Accessed: Apr. 28, 2024. [Online]. Available: https: //www.ncbi.nlm.nih.gov/books/NBK11770/

[6] H. N. Perry *et al.*, "Planning an integrated disease surveillance and response system: A matrix of skills and activities," *BMC Med.*, vol. 5, no. 1, pp. 1–8, Aug. 2007, doi:10.1186/1741-7015-5-24/FIGURES/3.

[7] S. Agrebi and A. Larbi, "Use of artificial intelligence in infectious diseases," *Artif. Intell. Precis. Heal.*, p. 415, Jan. 2020, doi:10.1016/B978-0-12-817133-2.00018-5.

[8] A. Bohr and K. Memarzadeh, "The rise of artificial intelligence in healthcare applications," *Artif. Intell. Healthc.*, pp. 25–60, Jan. 2020, doi:10.1016/B978-0-12-818438-7.00002-2.

[9] S. L. Groseclose and D. L. Buckeridge, "Public Health Surveillance Systems: Recent Advances in Their Use and Evaluation," *Annu. Rev. Public Health*, vol. 38, no. Volume 38, 2017, pp. 57–79, Mar. 2017, doi:10.1146/ANNUREV-PUBLHEALTH-031816-044348/CITE/REFWORKS.

[10] "The Emerging Science of Very Early Detection of Disease Outb... : Journal of Public Health Management and Practice." https: //journals.lww.com/jphmp/abstract/2001/07060/the_emerging_science_of_very_early_detection_of.6.aspx (accessed Apr. 29, 2024).

[11] C. Castillo-Salgado, "Trends and Directions of Global Public Health Surveillance," *Epidemiol. Rev.*, vol. 32, no. 1, pp. 93–109, Apr. 2010, doi:10.1093/EPIREV/MXQ008.

[12] N. Krieger *et al.*, "Assessing health impact assessment: multidisciplinary and international perspectives," *J. Epidemiol. Community Heal.*, vol. 57, no. 9, pp. 659–662, Sep. 2003, doi:10.1136/JECH.57.9.659.

[13] P. A. Rogerson and I. Yamada, "Monitoring change in spatial patterns of disease: comparing univariate and multivariate cumulative sum approaches," *Stat. Med.*, vol. 23, no. 14, pp. 2195–2214, Jul. 2004, doi:10.1002/SIM.1806.

[14] H. A. Shah and M. Househ, "Concepts, objectives and analysis of public health surveillance systems," *Comput. Methods Programs Biomed. Updat.*, vol. 5, p. 100136, Jan. 2024, doi:10.1016/J.CMPBUP.2024.100136.

[15] C. Li, G. Ye, Y. Jiang, Z. Wang, H. Yu, and M. Yang, "Artificial Intelligence in battling infectious diseases: A transformative role," *J. Med. Virol.*, vol. 96, no. 1, p. e29355, Jan. 2024, doi:10.1002/JMV.29355.

[16] "Global antimicrobial resistance forum launched to help tackle common threat to planetary health." https: //www.who.int/news-room/articles-detail/global-antimicrobial-resistance-forum-launched-to-help-tackle-common-threat-to-planetary-health (accessed Apr. 29, 2024).

[17] S. K. Ahmed *et al.*, "Antimicrobial resistance: Impacts, challenges, and future prospects," *J. Med. Surgery, Public Heal.*, vol. 2, p. 100081, Apr. 2024, doi:10.1016/J.GLMEDI.2024.100081.

[18] K. H. Luepke *et al.*, "Past, Present, and Future of Antibacterial Economics: Increasing Bacterial Resistance, Limited Antibiotic Pipeline, and Societal Implications," *Pharmacotherapy*, vol. 37, no. 1, pp. 71–84, Jan. 2017, doi:10.1002/PHAR.1868.

[19] R. Smith and J. Coast, "The true cost of antimicrobial resistance," *BMJ*, vol. 346, no. 7899, Mar. 2013, doi:10.1136/BMJ.F1493.

[20] P. D. Tamma and S. E. Cosgrove, "Antimicrobial Stewardship," *Infect. Dis. Clin. North Am.*, vol. 25, no. 1, pp. 245–260, Mar. 2011, doi:10.1016/j.idc.2010.11.011.

[21] "Division of Health Care Services: The future of public... - Google Scholar." https: //scholar.google.com/scholar_lookup?title=The Future of Public Health&publication_year=1988&author=Inst. Med.#d=gs_cit&t=1714453143904&u=%2Fscholar%3Fq%3Dinfo%3AKBhX7M8UHAQJ%3Ascholar.google.com%2F%26output%3Dcite%26scirp%3D0%26hl%3Den (accessed Apr. 30, 2024).

[22] S. Bhattacharya, H. Tran, P. Srinivasan, and J. Suls, "Belief surveillance with twitter," *Proc. 4th Annu. WebSci'12*, vol. volume, pp. 43–46, 2012, doi:10.1145/2380718.2380724.

[23] D. Zeng, Z. Cao, and D. B. Neill, "Artificial intelligence–enabled public health surveillance—from local detection to global epidemic monitoring and control," *Artif. Intell. Med.*, p. 437, Jan. 2021, doi:10.1016/B978-0-12-821259-2.00022-3.

[24] W. H. Organization, "Ethics and governance of artificial intelligence for health: WHO guidance: executive summary," 2021, Accessed: Apr. 30, 2024. [Online]. Available: https: //apps.who.int/iris/bitstream/handle/10665/350567/9789240037403-eng.pdf

[25] O. E. Santangelo, V. Gentile, S. Pizzo, D. Giordano, and F. Cedrone, "Machine Learning and Prediction of Infectious Diseases: A Systematic Review," *Mach. Learn. Knowl. Extr.*, vol. 5, no. 1, pp. 175–198, Mar. 2023, doi:10.3390/MAKE5010013/S1.

[26] J. A. Roth, M. Battegay, F. Juchler, J. E. Vogt, and A. F. Widmer, "Introduction to Machine Learning in Digital Healthcare Epidemiology," *Infect. Control Hosp. Epidemiol.*, vol. 39, no. 12, pp. 1457–1462, Dec. 2018, doi:10.1017/ICE.2018.265.

[27] O. Mesa Varona *et al.*, "Monitoring antimicrobial resistance and drug usage in the human and livestock sector and foodborne antimicrobial resistance in six European countries," *Taylor Fr. Mesa Varona, K Chain. B Muller-Pebody, MF Anjum, T Eckmanns, M NorströmInfection drug Resist. 2020•Taylor Fr.*, vol. 13, pp. 957–993, 2020, doi:10.2147/IDR.S237038.

[28] G. H. Ionescu, D. Firoiu, R. Pîrvu, M. Enescu, M. I. Rădoi, and T. M. Cojocaru, "The potential for innovation and entrepreneurship in eu countries in the context of sustainable development," *Sustain.*, vol. 12, no. 18, pp. 1–18, Sep. 2020, doi:10.3390/SU12187250.

[29]R. Sikkema, M. K.-I. E. & Epidemiology, and undefined 2016, "One Health training and research activities in Western Europe," *Taylor Fr. Sikk. M KoopmansInfection Ecol. Epidemiol. 2016•Taylor Fr.*, vol. 6, no. 1, p. 33703, 2016, doi:10.3402/iee.v6.33703.

[30] N. Macesic, F. Polubriaginof, and N. P. Tatonetti, "Machine learning: Novel bioinformatics approaches for combating antimicrobial resistance," *Curr. Opin. Infect. Dis.*, vol. 30, no. 6, pp. 511–517, Dec. 2017, doi:10.1097/QCO.0000000000000406.

[31] P. Bradley *et al.*, "Rapid antibiotic-resistance predictions from genome sequence data for Staphylococcus aureus and Mycobacterium tuberculosis," *Nat. Commun. 2015 61*, vol. 6, no. 1, pp. 1–15, Dec. 2015, doi:10.1038/ncomms10063.
[32] G. Arango-Argoty, E. Garner, A. Pruden, L. S. Heath, P. Vikesland, and L. Zhang, "DeepARG: A deep learning approach for predicting antibiotic resistance genes from metagenomic data," *Microbiome*, vol. 6, no. 1, pp. 1–15, Feb. 2018, doi:10.1186/S40168-018-0401-Z/FIGURES/9.
[33] M. T. E. Suller and A. D. Russell, "Triclosan and antibiotic resistance in Staphylococcus aureus," *J. Antimicrob. Chemother.*, vol. 46, no. 1, pp. 11–18, Jul. 2000, doi:10.1093/JAC/46.1.11.
[34] S. Suzuki, M. Ogo, T. Koike, H. Takada, and B. Newman, "Sulfonamide and tetracycline resistance genes in total- and culturable-bacterial assemblages in south african aquatic environments," *Front. Microbiol.*, vol. 6, no. AUG, p. 150078, Aug. 2015, doi:10.3389/FMICB.2015.00796/BIBTEX.
[35] M. Nguyen *et al.*, "Developing an in silico minimum inhibitory concentration panel test for Klebsiella pneumoniae," *Sci. Reports 2017 81*, vol. 8, no. 1, pp. 1–11, Jan. 2018, doi:10.1038/s41598-017-18972-w.
[36] A. Drouin *et al.*, "Predictive computational phenotyping and biomarker discovery using reference-free genome comparisons," *BMC Genomics*, vol. 17, no. 1, pp. 1–15, Sep. 2016, doi:10.1186/S12864-016-2889-6/FIGURES/5.
[37] J. J. Davis *et al.*, "Antimicrobial Resistance Prediction in PATRIC and RAST," *Sci. Reports 2016 61*, vol. 6, no. 1, pp. 1–12, Jun. 2016, doi:10.1038/srep27930.
[38] H. J. Lau, C. H. Lim, S. C. Foo, and H. S. Tan, "The role of artificial intelligence in the battle against antimicrobial-resistant bacteria," *Curr. Genet.*, vol. 67, no. 3, pp. 421–429, Jun. 2021, doi:10.1007/S00294-021-01156-5/METRICS.
[39] H. L. Her and Y. W. Wu, "A pan-genome-based machine learning approach for predicting antimicrobial resistance activities of the Escherichia coli strains," *Bioinformatics*, vol. 34, no. 13, pp. i89–i95, Jul. 2018, doi:10.1093/BIOINFORMATICS/BTY276.
[40] J. C. Hyun, E. S. Kavvas, J. M. Monk, and B. O. Palsson, "Machine learning with random subspace ensembles identifies antimicrobial resistance determinants from pan-genomes of three pathogens," *PLOS Comput. Biol.*, vol. 16, no. 3, p. e1007608, 2020, doi:10.1371/JOURNAL.PCBI.1007608.
[41] E. S. Kavvas *et al.*, "Machine learning and structural analysis of Mycobacterium tuberculosis pan-genome identifies genetic signatures of antibiotic resistance," *Nat. Commun. 2018 91*, vol. 9, no. 1, pp. 1–9, Oct. 2018, doi:10.1038/s41467-018-06634-y.
[42] R. M. Sherman and S. L. Salzberg, "Pan-genomics in the human genome era," *Nat. Rev. Genet. 2020 214*, vol. 21, no. 4, pp. 243–254, Feb. 2020, doi:10.1038/s41576-020-0210-7.
[43] J. Kim *et al.*, "VAMPr: VAriant Mapping and Prediction of antibiotic resistance via explainable features and machine learning," *PLOS Comput. Biol.*, vol. 16, no. 1, p. e1007511, 2020, doi:10.1371/JOURNAL.PCBI.1007511.
[44] A. A. Rabaan *et al.*, "Application of Artificial Intelligence in Combating High Antimicrobial Resistance Rates," *Antibiotics*, vol. 11, no. 6, Jun. 2022, doi:10.3390/ANTIBIOTICS11060784.

[45] J. Lv, S. Deng, L. Z.-B. and Health, and undefined 2021, "A review of artificial intelligence applications for antimicrobial resistance," *mednexus.orgJ Lv, S Deng, L ZhangBiosafety Heal. 2021•mednexus.org*, vol. 3, no. 1, pp. 22–31, Feb. 2020, doi:10.1016/j.bsheal.2020.08.003.

[46] I. Yelin *et al.*, "Personal clinical history predicts antibiotic resistance of urinary tract infections," *Nat. Med. 2019 257*, vol. 25, no. 7, pp. 1143–1152, Jul. 2019, doi:10.1038/s41591-019-0503-6.

3

Artificial Intelligence in Antibiotic Discovery and Development

Shyam Tripathi,[1,2] *Bhumika Chhoker,*[1,2]
Arpana Vibhuti[1,2] and *V. Samuel Raj*[1,2,*]

The integration of artificial intelligence (AI) and machine learning algorithms has revolutionized the discovery and development of new antibiotics by swiftly screening extensive chemical databases to pinpoint promising compounds, resulting in the identification of potent antibiotic candidates. This cutting-edge approach not only expedites the process but also fine-tunes existing antibiotic molecules to enhance their efficacy against specific pathogens. Through the fusion of AI with computational techniques like molecular dynamics simulations, a deeper understanding of antibiotic-bacteria interactions is achieved, streamlining the drug discovery pipeline. The emphasis on open science practices and interdisciplinary collaboration between computer scientists and microbiologists is pivotal to fully harnessing the potential of AI in antibiotic discovery, presenting a promising avenue to address the escalating challenge of antimicrobial resistance. AI has been instrumental in various facets of antibiotic discovery, from rapid screening of chemical databases and optimization of existing antibiotics to *de novo* molecular

[1] Department of Biotechnology, SRM University, Delhi-NCR, Rajiv Gandhi Education City, Sonepat - 131029, Haryana, India.
[2] Centre for Drug Design Discovery and Development (C4D), SRM University, Delhi-NCR, Rajiv Gandhi Education City, Sonepat - 131 029, Haryana, India.
* Corresponding author: samuelrajv@yahoo.com

design, antimicrobial resistance prediction, machine learning for efficacy prediction, deep learning for complex system analysis, natural product exploration, antibacterial systems biology, and drug combination development to combat multi-drug-resistant pathogens. These diverse applications of AI have significantly advanced the field of antibiotic discovery and deepened our comprehension of antimicrobial resistance, offering a hopeful solution to the pressing global issue.

3.1 Introduction

A natural phenomenon known as Antimicrobial Resistance (AMR) occurs when bacteria, viruses, fungi, and parasites develop the ability to resist the drugs that are intended to eradicate them. The development and spread of antimicrobial resistance has been made easier by the improper and overuse of antibiotics in agriculture, human healthcare, and the environment [1]. This alteration renders previously effective treatments useless, leading to prolonged illnesses and higher death rates that necessitate prompt and effective response. The global antimicrobial resistance and usage surveillance system, which was introduced by the World Health Organization (WHO), has shown that AMR is increasing and is already a significant cause of death worldwide [2,3].On a global scale, it is estimated that in 2019 only, roughly 4.95 million deaths were caused by bacterial antimicrobial resistance (AMR) with 1.27 million deaths directly related to bacterial AMR [2]. Based on findings from the Centers for Disease Control and Prevention (CDC) AMR to basic antibiotics causes about two million infections every year in the United states alone, resulting in at least 23,000 fatalities [4,5]. In 2019, the United States witnessed approximately 2.8 million infections caused by antibiotic resistant bacteria. AMR is expected to cause 10 million deaths per year by 2050. The Infectious Disease Society of America has identified six pathogens collectively known as "ESKAPE" species, that pose a serious threat to human health due to their fast development of antibiotic resistance. The pathogens are *Enterocuccus faecium, Staphylococcus aureus, Klebsiella pneumoniae, Acinetobacter baumannii, Pseudomonas aeruginosa and Enterobacter* spp. [5]. The World Health Organization (WHO) has released a list of antibiotic-resistant "priority pathogens" to help drug manufacturers prioritize urgent requirements for a few antibiotics. Furthermore. AMR has arisen as a major public health concern in China. The Chinese Antimicrobial surveillance network data show that the resistance rate of carbapenem-resistant Gram-negative bacteria has increased significantly. Interestingly, between 2005 and 2022, the prevalence of carbapenem-resistant A. baumannii increased from 39.0% to 71.9%, whereas carbapenem-resistant K. pneumoniae increased from 2.9% to 24.2% [6].

AMR is emerging as a major worldwide concern, demanding immediate action. To address this, the Food and Drug Administration (FDA) of the United States has suggested regulations outlining the particular types, amounts, and frequency of antibiotic usage that are considered suitable [7]. In 2006, the European Union proposed a broad ban on antibiotic usage in cow feed. Unlike their peers in Europe and the United States, officials in Japan and China have focused on more consequential initiatives. For example, in 2016, the Chinese government released the National Action Plan to Contain Antimicrobial Resistance (NAPACAR) [8]. Although public awareness of AMR has grown, the overall situation is deteriorating. To successfully address AMR, we must continue to produce antimicrobial peptides (AMPs), experiment with antibiotic combinations, and install strong monitoring systems [9]. The number of novel medicines produced and authorized has significantly dropped over the last decade, with just four new antibiotics licensed between 2010 and 2014 [10], resulting in restricted treatment choices for antibiotic-resistant bacterial infections in hospitals. Antibiotics were previously discovered by screening secondary metabolites from soil microorganisms that showed antibacterial activity [11]. However, the search for new antibiotics is becoming increasingly difficult, due in part to the "rediscovery problem," in which the same molecules are discovered repeatedly [12]. Hence, the current pace of new drug development is inadequate to address the need of clinical treatment, particularly for pathogens listed as priorities by the World Health Organization (WHO).

Artificial intelligence (AI), a subfield of computer science, involves creating intelligent machines capable of performing tasks that traditionally necessitate human-like intelligence in an unbiased manner [13]. AI technologies provide innovative approaches and have gradually infiltrated different disciplines to accelerate scientific achievements, particularly in medicine, where AI has aided in the identification of novel medications and expedited drug development and clinical research procedures [14,15,16,17]. Without fail, AI has played a critical role in joint multidisciplinary efforts to address the problem of antimicrobial resistance (AMR) [15]. AI technology is a powerful tool in combating AMR. AI can be utilized in various ways to address the challenges posed by AMR. Firstly, AI, particularly machine learning algorithms, can analyze vast datasets of microbial genetic information, clinical records, and drug interactions to predict the emergence of antibiotic-resistant strains and assist healthcare professionals in making informed decisions about antibiotic use [18]. Additionally, AI-powered diagnostics are revolutionizing the identification of infectious agents and their susceptibility to antibiotics, enabling rapid and accurate diagnosis crucial for effective treatment and reducing unnecessary antibiotic use [19]. Moreover, AI plays a pivotal role in drug discovery by accelerating the development of new antibiotics and alternative treatments through the prediction of potential antimicrobial compounds with higher efficiency. AI-powered surveillance systems continuously monitor antimicrobial usage, resistance patterns, and disease outbreaks, aiding in identifying regions or facilities with elevated AMR risks for targeted interventions to prevent its spread [20]. By integrating AI technologies into research, diagnosis, and treatment strategies, there is hope in the battle against AMR, emphasizing the importance of collaborative efforts that combine AI with traditional approaches to address this global health challenge effectively. I n recent years, AI has demonstrated great success in the treatment of antibiotic resistance. For example, AI tools based on sequencing techniques have been used to explore AMR [9]. Furthermore, aggregating clinical data to create a clinical decision support system might help healthcare professionals track antimicrobial resistance trends, allowing for more judicious antibiotic use [21].

The flowchart in Fig. 1 encapsulates the essential steps in the drug design and discovery process enhanced by AI, showcasing how AI technologies can revolutionize and optimize each stage for more efficient and effective drug development.

3.2 Role of Big Data and Artificial Intelligence in Drug Discovery and Development

Advancements in technology, paired with lower equipment costs have resulted in a rise in data creation both in number and diversity. This data flow has resulted in multiple data resources. Big data refers to datasets that are too large and complicated to be evaluated using conventional data analysis software, tools and procedures. It is characterized by three main features: volume, velocity and variety. Volume refers to the massive amount of data generated. Velocity represents the speed of data being produced. And the variety denotes the variability contained in datasets [22]. With the continual expansion of biological big data,

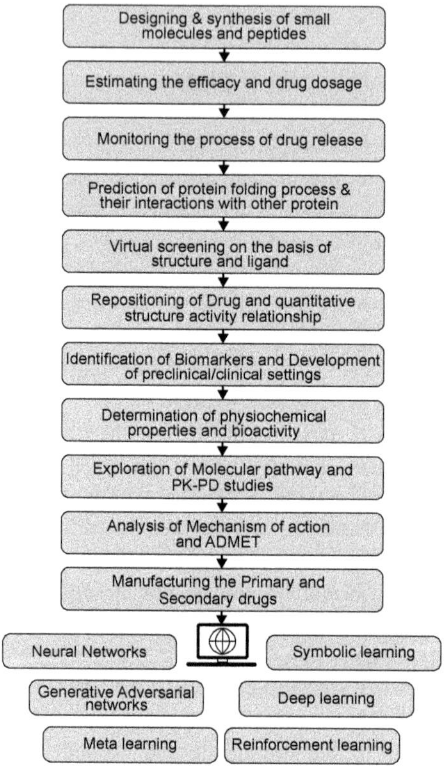

Figure 1: Flow chart to summarize the Drug Design and discovery process by using AI.

a multitude of artificial intelligence (AI) methods for analzying such data have surfaced [23]. AI technology empowers computers to learn and enhance autonomously devoid of explicit programming as well as to build and forecast models based on data. AI technique consists essentially of reasoning knowledge representation, solution finding and machine learning (ML) [24]. Deep Learning (DL), a subset of ML, uses neural networks to mimic the organization of the human brain. DL is used to find and classify patterns in language, photos, videos and many forms of biological data [25,26].

In recent years, DL algorithms have advanced rapidly. Several notable algorithms such as Convolutional Neural Networks (CNNs) Re current Neural Networks (RNNs) Deep Reinforcement Learning and most notably Generative Adversarial Networks (GANs) as an unsupervised learning algorithm have been widely used in various domains of drug discovery [27,28,29,30]. AI technology is an effective tool in the battle against an AMR [31,32,33,34]. For example, data-driven tools may be used to predict new antibiotic compounds whereas imaged based methods can help discover resistant bacteria. AI-powered compound library screening and novel compound structure design can speed up the discovery of more promising antibacterial drugs. Furthermore, AI may use current data such as genetic data to predict probable resistance locations and enzyme activities opening the path for the development of better antibiotics [35]. Moreover, AI has aided in target discovery and dynamic modelling, peptide design and synthesis, evaluation of structure- activity correlations (SAR) and structure-toxicity relationships (STR) and medication repurposing [33,36].

Additionally, big data applications extend to lead compound identification through screening chemical libraries and predicting drug-target interactions, optimizing lead compounds, enhancing preclinical studies by predicting drug safety and efficacy, identifying adverse effects, and optimizing drug formulations as shown in Fig. 2. Moreover, big data aids in improving clinical trials by enhancing patient recruitment and retention, identifying biomarkers, and monitoring drug safety and efficacy in real-time. Furthermore, big data streamlines the regulatory approval process, monitors drug safety and effectiveness, and identifies new indications or patient populations, ultimately making the drug discovery and development process more efficient, effective, and data-driven, leading to the development of safer and more effective drugs for patients.

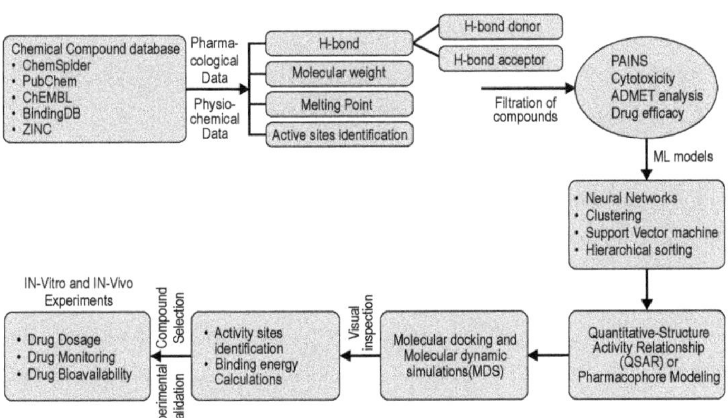

Figure 2: A pictorial representation of how big data can be leveraged in the drug discovery and development process.

The rapid growth of biological and chemical data from various sources such as literature, *in vitro* and *in vivo* studies, clinical trials, genomics, proteomics, metabolomics, gene ontology, and molecular pathway analysis has led to the development of numerous data repositories. These repositories, including Chemspider, ChEmbl, ZINC, BindingDB, and PubChem, serve as crucial tools for researchers involved in drug discovery and development. These databases simplify the drug design and development process by providing curated and screened data for pharmacological and physicochemical properties of compounds required for drug discovery. The high-throughput screened data undergo filtration based on drug-likeness, PAINS calculation, ADMET analysis, and toxicity. The filtered compounds are then analyzed further using deep learning (DL), random forest, classification, regression, and neural networks for further use. The compounds then undergo Quantitative Structure-Activity Relationship (QSAR) and pharmacophore modeling, followed by molecular docking and molecular dynamics simulation studies. The outcomes of the predicted compounds are calculated to determine binding energies and identify active sites. The final compound is validated using *in vitro* and *in vivo* experiments. In contrast, Quantum M echanical (QM) features play a significant role in drug discovery and development methods, such as ab initio density functional theory and semi-empirical computations, which use electron correlation methods to provide precise results. QM is expected to gain more importance in the arsenal of computational medicinal chemists, with modern QM methods having a greater direct influence on directing and improving the drug development process.

3.3 Role of AI in Improvement of Traditional Computational Drug Design

For years, computational tools have been essential in drug design and development, transforming the entire process. However, classic computational approaches continue to encounter obstacles such as time, computational cost and dependability [37,38] AI has the ability to overcome these limitations in virtual drug design. Therefore, expanding the importance of virtual approaches in drug discovery. Furthermore, the emergence of ML-based technologies has made it simpler to determine the 3-D structure of a protein of interest, which is a critical step in drug discovery. Novel medications are developing using the three-dimensional ligand binding environment of a protein [39,40] Google's DeepMind has created AlphaFold, an AI based tool trained on PDB structural data, to predict the 3D structure of proteins based on their amino acid sequences [41]. AlphaFold works in two steps: first it uses a convolutional neural network (CNN) to convert a protein's amino acid sequence into a distance matrix and a torsion angle matrix, and then it uses gradient optimization techniques to translate these matrices into the protein's three- dimensional structure [42]. Similarly, Mohammed AL Quraishi of Harvard Medical School created Recurrent Geometric Network (https: //github. com/a laboratory/run), a deep learning program that creates a protein's three-dimensional structure from its amino acid sequence. This model uses a single neural network to calculate bond angles and rotations of chemical bonds joining distinct amino acids, allowing for the prediction of a protein's 3D-structure [43]. Furthermore, quantum mechanics is used to determine the characteristics of molecules at a subatomic scale which is helpful in the prediction of protein-ligand interactions during drug development. At times, traditional computational approaches struggle with the computational needs of quantum physics leading to significant expense and complexity. This computing cost might affect the accuracy of the results [44]. Certainly AI has the potential to make quantum physics more approachable and useful. In drug development molecular dynamics (MD) simulations are critical for assessing protein- ligand interactions and binding stability. However, MD simulation may be tedious and time consuming. Ai has the potential to speed the MD simulation process, overcoming this difficulty [45]. In addition, de novo drug creation has incorporated AI in recent years. For example, Q. Bai et al. (2020) created MolAIcal (https: //molaical.github.io/), a platform for building three-dimensional drugs in protein pockets [46]. MolAICal utilizes deep learning and genetic algorithms trained on FDA-approved pharmaceuticals for de novo drug creation, and combines molecular docking with a deep learning model trained on the ZINC database for 3D drug design [47]. Similarly, Popova et al. (2018) developed ReLeaSE (https: //github.com/isayev/ReLeaSE), a deep reinforcement learning approach for de novo drug creation. ReLeaSE combines two deep neural networks (DNNs): generative and predictive, with the generative model creating novel compounds and the predictive model forecasting the attributes of the molecule [48].

3.4 Artificial Intelligence in Primary and Secondary Drug Screening

AI is currently a highly effective and desirable technology due to its ability to save time and money [49]. AI technology plays an essential role in initial drug screening as shown in Fig. 3, since it helps to classify and arrange cells using image analysis. Although many machine learning (ML) models using various methods achieve impressive accuracy in picture identification they frequently encounter difficulties when dealing with big datasets. To correctly categorize the target cell, the ML model must be trained to recognize the cell and its characteristics. This training often entails contrasting the picture of the targeted cells to identify them from the background [50]. Various textured characteristics are retrieved from the photos, including wavelet-based texture features and Tamura texture features, and their dimensions are then decreased using principal component analysis (PCA). According to research, the least-square SVM(LS-SVM) achieved the maximum classification accuracy of 95.34% [47,48]. In terms of cell sorting, speed is important for efficiently isolating the target cell type from a sample. According to the research, image-activated cell sorting (IACS) is at the cutting edge of this field, capable of evaluating cells optical, electrical and mechanical properties [51]. Secondary drug screening assesses a substance's physical characteristics, bioactivity and toxicity. Physical parameters such as melting point and partition coefficient are important factors in influencing a compound's bioavailability and should be taken into account when designing compounds [27]. Molecular representation approaches such as molecular fingerprinting, simplified molecular-input line entry system (SMILES) and coulomb matrices are used in drug design [52,53]. This data may be efficiently used in Deep N eural N etworks (DNN), which have two separate stages: generative and predictive. While both the stages are trained separately using supervised learning, their combined training might inject bias into the output by rewarding or punishing it for particular attributes. This complete procedure may be used for reinforcement learning [54,55]. Matched Molecular Pair (MMP) analysis is commonly used in QSAR (Quantitative Structure-Activity Relationship) research. MMP examines a single change in a therapeutic candidate that affects the compound bioactivity [53]. In addition to MMP, additional machine learning techniques such as Random Forest (RF) and Gradient Boosting Machines (GBM) are used to create changes. Studies show that DNN beats RF and GBM in prediction accuracy [56]. The development of publicly accessible databases like as ChEMBL, PubChem and ZINC has given us access to vast libraries of compounds each annotated with information such as structure, known targets and availability for purchase. Combining Matched Molecular Pair (MMP) analysis with ML approaches allows us to anticipate bioactivity characteristics such as oral exposure intrinsic clearance, ADMET parameters and mechanism of action [57].

In the drug development and design pipeline, the screening of potential leads is a crucial step, with artificial intelligence (AI) playing a pivotal role in identifying unique and promising lead molecules. The vast chemical space, encompassing over 106 million chemical structures derived from various research sources like OMIC investigations, clinical trials, *in vivo* experiments, and microarray analysis, is screened using machine learning algorithms such as reinforcement learning, logistic regression, ordinary regression, and generative models. These algorithms are employed to assess the chemical compounds for active

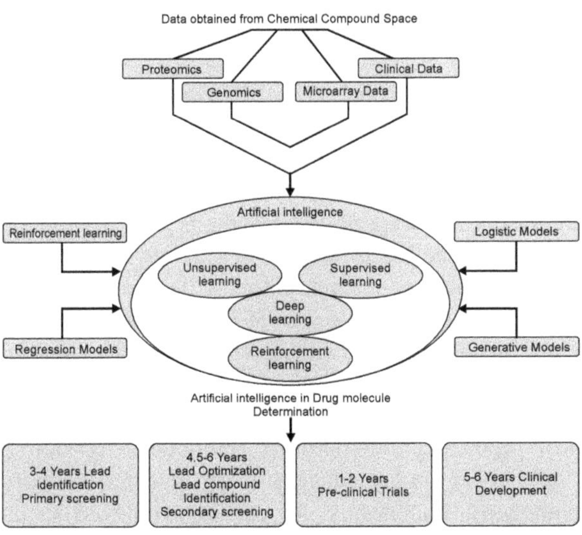

Figure 3: The integration of Artificial Intelligence (AI) has revolutionized the primary and secondary screening processes in both medical and drug discovery domains. In primary medical screening, AI-powered algorithms are employed to accurately and sensitively identify and classify cells, such as cancer cells, in medical images, enabling early detection and monitoring of diseases. Conversely, in the drug discovery process, AI accelerates the identification and testing of new drugs by predicting how different chemical compounds will interact with biological targets during secondary screening. This streamlines the drug development process and improves the success rate of identifying effective compounds, ultimately leading to more efficient and effective drug discovery.

sites, structural properties, and their ability to bind to specific targets. The drug development process supported by AI typically lasts 14 to 18 years, significantly shorter than traditional methods. The initial phase involves identifying disease-modifying target proteins through technologies like reverse docking, bioinformatics analysis, and computational chemical biology. In the subsequent phase, compounds are screened to identify potential lead candidates capable of inhibiting the target protein, utilizing virtual screening and de novo design. Leads are further optimized and selected through targeted library design, drug-like analysis, drug-target reproducibility evaluation, and computational biology techniques. Subsequently, compounds undergo secondary screening before progressing to pre-clinical studies. The final phase, clinical development, involves cell-culture analysis, animal model experiments, and patient analysis, culminating in the comprehensive evaluation of the drug's efficacy and safety profile.

3.5 Artificial Intelligence Applications in Accelerating Antibiotic Discovery

Artificial intelligence (AI) has been extensively applied to accelerate the discovery of new antibiotics through various innovative approaches. AI algorithms can rapidly screen large chemical databases to identify promising antibiotic compounds, a process that is much faster and more efficient than traditional experimental methods [58]. Additionally, AI can be utilized to optimize the properties of existing antibiotic molecules, such as improving their potency or targeting them to specific pathogens, further enhancing their effectiveness [59]. The power of AI also extends to de novo molecular design, where it can generate novel molecular structures with antimicrobial properties, potentially leading to the discovery of new antibiotics. AI can further assist in the identification of the most promising antimicrobial compounds by analyzing their chemical properties and predicting their potential efficacy against various pathogens. Importantly, AI can be employed to predict the likelihood of antimicrobial resistance developing against a particular antibiotic, aiding in the selection of more effective treatments. Machine learning algorithms trained on large datasets of known antibiotics and their properties can also be leveraged to predict the potential efficacy of new compounds against specific pathogens [59,20]. Deep learning techniques have been applied to analyze complex biological systems and predict the interactions between antibiotics and bacteria, leading to more effective and targeted treatments [59]. AI has also shown promise in aiding the exploration of natural products for antimicrobial properties and analyzing the complex interactions between bacteria and their environments to identify potential targets for new antibiotics [60]. Finally, AI can be applied to identify optimal combinations of existing antibiotics to combat multi drug-resistant pathogens. These diverse applications of AI have significantly accelerated the discovery of new antibiotics and improved our understanding of antimicrobial resistance [58,59,60].

3.6 Conclusion

The integration of artificial intelligence (AI) in antibiotic discovery represents a ground-breaking advancement in combating antimicrobial resistance (AMR). AI technologies have revolutionized the process by rapidly screening chemical databases, optimizing existing antibiotic molecules, and even designing novel molecular structures with antimicrobial properties. The predictive capabilities of AI in identifying promising antimicrobial compounds, predicting resistance development, and exploring natural products for new antibiotics have significantly enhanced the efficiency and effectiveness of antibiotic discovery. Moreover, AI's application in analyz ing complex biological systems, developing drug combinations, and exploring antibacterial systems biology has provided invaluable insights into combating multi-drug-resistant pathogens and understanding the intricate interactions between bacteria and their environments. The collaborative efforts between AI technologies and traditional drug discovery methods offer a promising solution to address the escalating global crisis of antimicrobial resistance, emphasizing the critical role of AI in reshaping the landscape of antibiotic development and fostering innovative approaches to combat AMR effectively.

References

[1] "Antibiotic resistance threats in the United States, 2019," Atlanta, Georgia, Nov. 2019. doi:10.15620/cdc: 82532.

[2] C. J. Murray *et al.*, "Global burden of bacterial antimicrobial resistance in 2019: a systematic analysis," *The Lancet*, vol. 399, no. 10325, pp. 629–655, Feb. 2022, doi:10.1016/S0140-6736(21)02724-0.

[3] Ajulo, S. and Awosile, B. 2024. Global antimicrobial resistance and use surveillance system (GLASS 2022): Investigating the relationship between antimicrobial

resistance and antimicrobial consumption data across the participating countries. PloS one, 19(2), e0297921. https://doi.org/10.1371/journal.pone.0297921

[4] "Antibiotic resistance threats in the United States, 2019," Atlanta, Georgia, Nov. 2019. doi:10.15620/cdc: 82532.

[5] J. N. Pendleton, S. P. Gorman, and B. F. Gilmore, "Clinical relevance of the ESKAPE pathogens," *Expert Review of Anti-Infective Therapy*, vol. 11, no. 3. pp. 297–308, 2013. doi:10.1586/eri.13.12.

[6] S. R. Chowdhury, Z. Ahamed, K. Roy, A. Al Noman, R. M. Haroon, and C. Mondol, "African Journal of Bacteriology Research Emerging threats of antibiotic resistance in Salmonella typhi and Salmonella paratyphi A among enteric fever cases of Dhaka, Bangladesh," vol. 14, no. 1, pp. 8–15, doi:10.5897/JBR2021.0340.

[7] T. Van Gemert, "On the influence of dataset characteristics on classifier performance."

[8] N. Ahmed, B. Zeshan, M. Naveed, and M. Afzal, "Antibiotic resistance profile in relation to virulence genes fimH, hlyA and usp of uropathogenic E. coli isolates in Lahore, Pakistan." [Online]. Available: https: //www.researchgate.net/publication/336555234

[9] J. Lv, S. Deng, and L. Zhang, "A review of artificial intelligence applications for antimicrobial resistance," *Biosafety and Health*, vol. 3, no. 1. Elsevier B.V., pp. 22–31, Feb. 01, 2021. doi:10.1016/j.bsheal.2020.08.003.

[10] C. Lee Ventola, "The Antibiotic Resistance Crisis Part 1: Causes and Threats," 2015.

[11] M. Hutchings, A. Truman, and B. Wilkinson, "Antibiotics: past, present and future," *Current Opinion in Microbiology*, vol. 51. Elsevier Ltd, pp. 72–80, Oct. 01, 2019. doi:10.1016/j.mib.2019.10.008.

[12] L. Katz and R. H. Baltz, "Natural product discovery: past, present, and future," *Journal of Industrial Microbiology and Biotechnology*, vol. 43, no. 2–3. Springer Verlag, pp. 155–176, Mar. 01, 2016. doi:10.1007/s10295-015-1723-5.

[13] J. G. Greener, S. M. Kandathil, L. Moffat, and D. T. Jones, "A guide to machine learning for biologists," *Nature Reviews Molecular Cell Biology*, vol. 23, no. 1. Nature Research, pp. 40–55, Jan. 01, 2022. doi:10.1038/s41580-021-00407-0.

[14] M. W. Mullowney *et al.*, "Artificial intelligence for natural product drug discovery," *Nature Reviews Drug Discovery*, vol. 22, no. 11. Nature Research, pp. 895–916, Nov. 01, 2023. doi:10.1038/s41573-023-00774-7.

[15] H. Wang *et al.*, "Scientific discovery in the age of artificial intelligence," *Nature*, vol. 620, no. 7972. Nature Research, pp. 47–60, Aug. 03, 2023. doi:10.1038/s41586-023-06221-2.

[16] E. J. Topol, "High-performance medicine: the convergence of human and artificial intelligence," *Nature Medicine*, vol. 25, no. 1. Nature Publishing Group, pp. 44–56, Jan. 01, 2019. doi:10.1038/s41591-018-0300-7.

[17] F. Wong, C. de la Fuente-Nunez, and J. J. Collins, "Leveraging artificial intelligence in the fight against infectious diseases," *Science*, vol. 381, no. 6654. American Association for the Advancement of Science, pp. 164–170, Jul. 14, 2023. doi:10.1126/science.adh1114.

[18] R. Masud, N. Fahim, M. Rana, M. Islam, and M. Rahman, "Artificial intelligence, a powerful tool to combat antimicrobial resis tance: An update," *Journal of Advanced Biotechnology and Experimental Therapeutics*, vol. 6, no. 3, p. 711, doi:10.5455/jabet.2023.d161.

[19] A. A. Rabaan *et al.*, "Application of Artificial Intelligence in Combating High Antimicrobial Resistance Rates.," *Antibiotics (Basel)*, vol. 11, no. 6, Jun. 2022, doi:10.3390/antibiotics11060784.

[20] T. Ali, S. Ahmed, and M. Aslam, "Artificial Intelligence for Antimicrobial Resistance Prediction: Challenges and Opportunities towards Practical Implementation.," *Antibiotics (Basel)*, vol. 12, no. 3, Mar. 2023, doi:10.3390/antibiotics12030523.

[21] H. J. Lau, C. H. Lim, S. C. Foo, and H. S. Tan, "The role of artificial intelligence in the battle against antimicrobial-resistant bacteria," *Current Genetics*, vol. 67, no. 3. Springer Science and Business Media Deutschland GmbH, pp. 421–429, Jun. 01, 2021. doi:10.1007/s00294-021-01156-5.

[22] T. Ching *et al.*, "Opportunities and obstacles for deep learning in biology and medicine," *J R Soc Interface*, vol. 15, no. 141, 2018, doi:10.1098/rsif.2017.0387.

[23] D. M. Camacho, K. M. Collins, R. K. Powers, J. C. Costello, and J. J. Collins, "Next-Generation Machine Learning for Biological Networks," *Cell*, vol. 173, no. 7. Cell Press, pp. 1581–1592, Jun. 14, 2018. doi:10.1016/j.cell.2018.05.015.

[24] Y. Lecun, Y. Bengio, and G. Hinton, "Deep learning," *Nature*, vol. 521, no. 7553. Nature Publishing Group, pp. 436–444, May 27, 2015. doi:10.1038/nature14539.

[25] J. Zou, M. Huss, A. Abid, P. Mohammadi, A. Torkamani, and A. Telenti, "A primer on deep learning in genomics," *Nat Genet*, vol. 51, no. 1, pp. 12–18, Jan. 2019, doi:10.1038/s41588-018-0295-5.

[26] I. J. Goodfellow *et al.*, "Generative Adversarial Networks," Jun. 2014, [Online]. Available: http: //arxiv.org/abs/1406.2661

[27] H. C. S. Chan, H. Shan, T. Dahoun, H. Vogel, and S. Yuan, "Advancing Drug Discovery via Artificial Intelligence," *Trends in Pharmacological Sciences*, vol. 40, no. 8. Elsevier Ltd, pp. 592–604, Aug. 01, 2019. doi:10.1016/j.tips.2019.06.004.

[28] G. Parkhe[1] and T. Thange[2], "A REVIEW: ARTIFICIAL INTELLIGENCE IN DRUG DISCOVERY AND DEVELOPMENT," 2022. [Online]. Available: www.ijrpr.com

[29] B. Digital, R. Belmont, D. Repository, and T. Valeriano, "'The Role of Artificial Intelligence in the Pharmaceutical Field: "The Role of Artificial Intelligence in the Pharmaceutical Field: Enhancing Therapeutic Outcomes and Repurposing through the Enhancing Therapeutic Outcomes and Repurposing through the Acceleration of Drug Discovery" Acceleration of Drug Discovery.'" [Online]. Available: https: //repository.belmont.edu/burs

[30] K. K. Mak and M. R. Pichika, "Artificial intelligence in drug development: present status and future prospects," *Drug Discovery Today*, vol. 24, no. 3. Elsevier Ltd, pp. 773–780, Mar. 01, 2019. doi:10.1016/j.drudis.2018.11.014.

[31] M. Miethke *et al.*, "Towards the sustainable discovery and development of new antibiotics," *Nature Reviews Chemistry*, vol. 5, no. 10. Nature Research, pp. 726–749, Oct. 01, 2021. doi:10.1038/s41570-021-00313-1.

[32] G. Liu, J. M. Stokes, D. Rogers, and R. E. Lee, "A brief guide to machine learning for antibiotic discovery," *ElsevierG Liu, JM StokesCurrent Opinion in Microbiology, 2022•Elsevier*, 2022, doi:10.1016/j.mib.2022.102190.

[33] F. Urbina, A. C. Puhl, and S. Ekins, "Recent advances in drug repurposing using machine learning," *Current Opinion in Chemical Biology*, vol. 65. Elsevier Ltd, pp. 74–84, Dec. 01, 2021. doi:10.1016/j.cbpa.2021.06.001.

[34] "Advisory Board and Contents," *Trends Cell Biol*, vol. 33, no. 7, p. i, Jul. 2023, doi:10.1016/s0962-8924(23)00094-6.

[35] A. Talat, A. K.-D. D. Today, and undefined 2023, "Artificial intelligence as a smart approach to develop antimicrobial drug molecules: A paradigm to combat drug-resistant infections," *ElsevierA Talat, AU KhanDrug Discovery Today, 2023•Elsevier*, Accessed: May 07, 2024. [Online]. Available: https: //www.sciencedirect.com/science/article/pii/S1359644623000077

[36] F. Urbina, A. C. Puhl, and S. Ekins, "Recent advances in drug repurposing using machine learning," *Current Opinion in Chemical Biology*, vol. 65. Elsevier Ltd, pp. 74–84, Dec. 01, 2021. doi:10.1016/j.cbpa.2021.06.001.

[37] A. Baldi, "Computational approaches for drug design and discovery: An overview," *Systematic Reviews in Pharmacy*, vol. 1, no. 1. pp. 99–105, Jan. 2010. doi:10.4103/0975-8453.59519.

[38] A. Lavecchia and C. Cerchia, "In silico methods to address polypharmacology: Current status, applications and future perspectives," *Drug Discovery Today*, vol. 21, no. 2. Elsevier Ltd, pp. 288–298, 2016. doi:10.1016/j.drudis.2015.12.007.

[39] J. S. Smith, A. E. Roitberg, and O. Isayev, "Transforming Computational Drug Discovery with Machine Learning and AI," *ACS Medicinal Chemistry Letters*, vol. 9, no. 11. American Chemical Society, pp. 1065–1069, Nov. 08, 2018. doi:10.1021/acsmedchemlett.8b00437.

[40] Y. Jing, Y. Bian, Z. Hu, L. Wang, and X. Q. Xie, "Correction to: Deep Learning for Drug Design: an Artificial Intelligence Paradigm for Drug Discovery in the Big Data Era (The AAPS Journal, (2018), 20, 3, (58), 10.1208/s12248-018-0210-0)," *AAPS Journal*, vol. 20, no. 4. Springer New York LLC, Jul. 01, 2018. doi:10.1208/s12248-018-0243-4.

[41] J. Powles and H. Hodson, "Google DeepMind and healthcare in an age of algorithms," *Health Technol (Berl)*, vol. 7, no. 4, pp. 351–367, Dec. 2017, doi:10.1007/s12553-017-0179-1.

[42] A. W. Senior *et al.*, "Improved protein structure prediction using potentials from deep learning," *Nature*, vol. 577, no. 7792, pp. 706–710, Jan. 2020, doi:10.1038/s41586-019-1923-7.

[43] M. AlQuraishi, "End-to-End Differentiable Learning of Protein Structure," *Cell Syst*, vol. 8, no. 4, pp. 292-301.e3, Apr. 2019, doi:10.1016/j.cels.2019.03.006.

[44] C. Kalaiarasi, S. Manjula, and P. Kumaradhas, "Combined quantum mechanics/molecular mechanics (QM/MM) methods to understand the charge density distribution of estrogens in the active site of estrogen receptors," *RSC Adv*, vol. 9, no. 69, pp. 40758–40771, 2019, doi:10.1039/c9ra08607b.

[45] M. De Vivo, M. Masetti, G. Bottegoni, and A. Cavalli, "Role of Molecular Dynamics and Related Methods in Drug Discovery," *Journal of Medicinal Chemistry*, vol. 59, no. 9. American Chemical Society, pp. 4035–4061, May 12, 2016. doi:10.1021/acs.jmedchem.5b01684.

[46] Q. Bai, S. Tan, T. Xu, H. Liu, J. Huang, and X. Yao, "MolAICal: A soft tool for 3D drug design of protein targets by artificial intelligence and classical algorithm," *Brief Bioinform*, vol. 22, no. 3, May 2021, doi:10.1093/bib/bbaa161.

[47] T. Sterling and J. J. Irwin, "ZINC 15 - Ligand Discovery for Everyone," *J Chem Inf Model*, vol. 55, no. 11, pp. 2324–2337, Nov. 2015, doi:10.1021/acs.jcim.5b00559.

[48] M. Popova, O. Isayev, and A. Tropsha, "Deep reinforcement learning for de novo drug design," 2018. [Online]. Available: https: //www.science.org

[49] T. H. Davenport, R. Ronanki, J. Wheaton, and A. Nguyen, "Feature Artificial Intelligence For The Real World 108 Harvard Business Review."

[50] R. K. Tripathy, S. Mahanta, and S. Paul, "Artificial intelligence-based classification of breast cancer using cellular images," *RSC Adv*, vol. 4, no. 18, pp. 9349–9355, 2014, doi:10.1039/c3ra47489e.
[51] P. Samui and D. P. Kothari, "Utilization of a Least Square Support Vector Machine (LSSVM) for slope stability analysis," *Scientia Iranica*, vol. 18, no. 1 A, pp. 53–58, 2011, doi:10.1016/j.scient.2011.03.007.
[52] C. W. L. Ho, D. Soon, K. Caals, and J. Kapur, "Governance of automated image analysis and artificial intelligence analytics in healthcare," *Clinical Radiology*, vol. 74, no. 5. W.B. Saunders Ltd, pp. 329–337, May 01, 2019. doi:10.1016/j.crad.2019.02.005.
[53] T. Andrysek, "Impact of physical properties of formulations on bioavailability of active substance: Current and novel drugs with cyclosporine," in *Molecular Immunology*, Elsevier Ltd, 2003, pp. 1061–1065. doi:10.1016/S0161-5890(03)00077-4.
[54] D. C. Elton, Z. Boukouvalas, M. S. Butrico, M. D. Fuge, and P. W. Chung, "Applying machine learning techniques to predict the properties of energetic materials," *Sci Rep*, vol. 8, no. 1, Dec. 2018, doi:10.1038/s41598-018-27344-x.
[55] M. Popova, O. Isayev, and A. Tropsha, "Deep reinforcement learning for de novo drug design," 2018. [Online]. Available: http: //advances.sciencemag.org/
[56] C. Tyrchan and E. Evertsson, "Matched Molecular Pair Analysis in Short: Algorithms, Applications and Limitations," *Computational and Structural Biotechnology Journal*, vol. 15. Elsevier B.V., pp. 86–90, 2017. doi:10.1016/j.csbj.2016.12.003.
[57] S. Turk, B. Merget, F. Rippmann, and S. Fulle, "Coupling Matched Molecular Pairs with Machine Learning for Virtual Compound Optimization," *J Chem Inf Model*, vol. 57, no. 12, pp. 3079–3085, Dec. 2017, doi:10.1021/acs.jcim.7b00298.
[58] T. Sterling and J. J. Irwin, "ZINC 15 –Ligand Discovery for Everyone," *J Chem Inf Model*, vol. 55, no. 11, pp. 2324–2337, Nov. 2015, doi:10.1021/acs.jcim.5b00559.
[59] M. C. R. Melo, J. R. M. A. Maasch, and C. de la Fuente-Nunez, "Accelerating antibiotic discovery through artificial intelligence," *Commun Biol*, vol. 4, no. 1, p. 1050, 2021, doi:10.1038/s42003-021-02586-0.
[60] F. Wong *et al.*, "Discovery of a structural class of antibiotics with explainable deep learning," *Nature*, vol. 626, no. 7997, pp. 177–185, 2024, doi:10.1038/s41586-023-06887-8.

4

Artificial Intelligence for Precision Medicine in Antimicrobial Resistance

Riya Mukherjee,[1] *Ramendra Pati Pandey*[2] and *Chung-Ming Chang*[1,3,4,*]

Artificial Intelligence for Precision Medicine in Antimicrobial Resistance" explores the transformative role of AI in combating antimicrobial resistance (AMR). It highlights how machine learning and data analytics can identify resistant pathogens, predict treatment outcomes, and personalize antibiotic therapies. By integrating AI with genomic and clinical data, the chapter illustrates advancements in diagnostic accuracy and treatment precision, reducing the misuse of antibiotics. It also addresses ethical considerations and the need for interdisciplinary collaboration. This chapter provides a comprehensive overview of AI's potential to revolutionize AMR management, ultimately improving patient outcomes and public health.

[1] Graduate Institute of Biomedical Sciences, Chang Gung University; No.259, Wenhua 1st Rd., Guishan Dist. Taoyuan City 33302, Taiwan (R.O.C.).
Email: d1001406@cgu.edu.tw; riya.mukherjee1896@gmail.com

[2] Biotechnology & Microbiology, SRM University Delhi-NCR, Sonepat, Rajiv Gandhi Education City, Sonepat - 131 029, Haryana.

[3] Department of Medical Biotechnology and Laboratory Science, Chang Gung University; No.259, Wenhua 1st Rd., Guishan Dist. Taoyuan City 33302, Taiwan (R.O.C.).

[4] Master & Ph.D. Program in Biotechnology Industry, Chang Gung University; No.259, Wenhua 1st Rd., Guishan Dist. Taoyuan City 33302, Taiwan (R.O.C.).

* Corresponding author

4.1 Introduction

Antimicrobial Resistance (AMR) represents a significant global health challenge, threatening the effectiveness of antibiotics and other antimicrobial agents used to treat infections. Over the past decades, the misuse and overuse of antimicrobials in human health, agriculture, and veterinary medicine have accelerated the development of AMR, leading to a looming crisis in healthcare systems worldwide. As pathogens evolve and become increasingly resistant to available treatments, the need for innovative solutions to combat AMR has become more urgent than ever. Precision medicine, a paradigm shift in healthcare, offers a promising approach to address AMR's challenges. Unlike the traditional one-size-fits-all approach to medical treatment, precision medicine emphasizes the customization of healthcare interventions based on individual variability in genes, environment, and lifestyle. By tailoring treatments to the specific characteristics of patients and pathogens, precision medicine can potentially optimize therapeutic outcomes while minimizing the risk of AMR development and spread.

Central to the advancement of precision medicine in the fight against AMR is the integration of A rtificial I ntelligence (AI) technologies into healthcare systems. Precision medicine and artificial intelligence (AI) together have the potential to completely transform the medical field. Precision medicine techniques pinpoint the characteristics of patients with uncommon treatment outcomes or distinct medical requirements. Artificial intelligence (AI) uses complex calculations and inference to produce insights, gives the system the ability to reason and learn, and enhances human decision-making in clinicians. AI, characterized by its ability to analyze vast amounts of data, identify patterns, and generate actionable insights, has emerged as a transformative force in various domains, including medicine. In the context of AMR, AI-powered tools and algorithms offer unprecedented opportunities to enhance the understanding of microbial dynamics, predict antimicrobial resistance patterns, and guide clinical decision-making.

This chapter explores the intersection of artificial intelligence, precision medicine, and antimicrobial resistance, highlighting the potential of AI-driven approaches to revolutionize the prevention, diagnosis, and treatment of antimicrobial-resistant infections. By leveraging the synergies between these advanced technologies, we aim to pave the way toward more effective strategies for combating AMR and safeguarding global public health.

4.2 Understanding of Antimicrobial Resistance

The ability of microorganisms, including bacteria, viruses, fungi, and parasites, to withstand the effects of drugs that were once successful against them is known as Antimicrobial Resistance (AMR). Genetic mutations or the acquisition of resistance genes from other microbes are the main ways that this resistance manifests itself. The misuse of antibiotics in human medicine, agriculture, and livestock, as well as inadequate sanitary conditions and infection control procedures, all play a role in the spread of antimicrobial resistance (AMR). AMR has serious side effects that include increased mortality, and higher medical expenses. AMR poses a danger to patient outcomes and the overall effectiveness of the healthcare system as infections grow more difficult to cure. This could eventually result in a situation where common infections and mild injuries could become fatal once more. This section explores the mechanisms, contributing factors, and consequences of AMR on patient outcomes and healthcare systems.

4.3 Mechanisms of Antimicrobial Resistance

Antibiotics have undoubtedly shielded the human population from many harmful bacterial infections, yet the phenomenon of antibiotic resistance is not new. Instances of penicillin resistance were reported as early as the mid-1990s [1]. Over time, the emergence of extensively drug-resistant (XDR), multi-drug-resistant (MDR), and more recently, Pan-Drug-Resistant (PDR) bacteria, have significantly increased the prevalence of resistance, especially in less developed countries. PDR bacteria exhibit resistance to all classes of antimicrobials, XDR bacteria are resistant to all but two categories, and MDR bacteria are resistant to at least one.

agent in three or more categories [2]. Understanding the mechanisms through which antibiotics function is crucial to addressing the mechanisms of resistance (Fig. 1). Generally, antibiotics work in one of five ways: (1) interfering with enzymes necessary for peptidoglycan synthesis, crucial for bacterial cell wall construction; (2) inhibiting nucleic acid synthesis; (3) obstructing protein synthesis; (4) disrupting metabolic pathways; and (5) damaging the cytoplasmic membrane. Each antibiotic may act through one or multiple of these mechanisms [2].

Microorganisms have evolved various metabolic pathways to develop resistance to these antimicrobial drugs[3]. Although these pathways are diverse, they can be categorized based on the frequency and medical

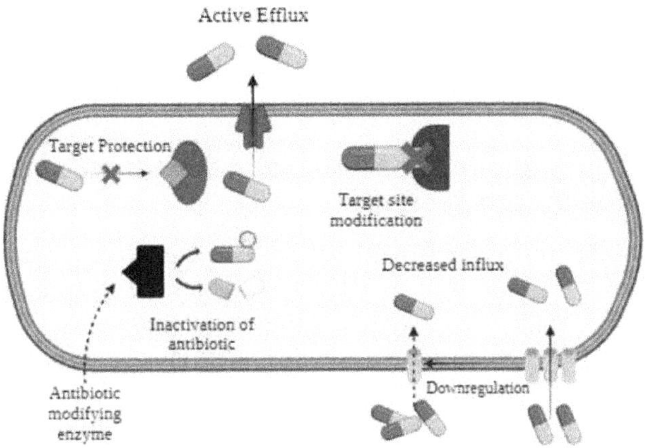

Figure 1: Different ways of mechanisms of antimicrobial resistance. It depicts the different mechanisms by which bacteria develop antimicrobial resistance. It illustrates the following processes within a bacterial cell: active efflux of antibiotics out of the cell, target site modification to prevent antibiotic binding, decreased influx by downregulating porins, antibiotic inactivation by modifying enzymes, and target protection where the bacterial proteins prevent the antibiotic from binding to its target. Each mechanism is depicted with arrows and labeled for clarity.

significance of the infections they cause. This understanding is essential for effectively developing strategies to combat antibiotic resistance [4.5].

4.4 Factors Contributing to the Spread of AMR

Human-to-human contact in healthcare facilities and the community is one of the main ways that antimicrobial resistance (AMR) is acquired and transmitted [4,5]. Antimicrobial-resistance genes can also be transported between and within reservoirs, which include humans, animals, water, and environmental sources. The bacterial species and resistance mechanisms involved can have a substantial impact on the specific transmission routes. A variety of areas, such as wastewater and sludge from urban wastewater treatment facilities and natural fertilizers including pig slurry, cow manure, and poultry farming fertilizer, serve as hotspots for the emergence of antimicrobial-resistant bacteria. The proliferation of resistance is significantly facilitated by these sources. Moreover, the primary mechanism for the spread of antibiotic resistance from animals to humans is the use of antibiotics in livestock feed and the consumption of these animals by humans [6]. Ingestion of fecal-contaminated food or drink as well as direct human-animal contact are other frequent routes of transmission. These paths emphasize the need for comprehensive methods to manage and mitigate AMR by stressing the interdependence of human, animal, and environmental health [7,8].

4.4.1 Consequences of AMR for patient outcomes and healthcare systems

Antimicrobial resistance (AMR) has far-reaching effects on the overall healthcare system as well as patient outcomes. As typical antibiotic treatments lose their effectiveness, antimicrobial resistance (AMR) can cause patients to become more susceptible to infections, prolong their illness, have higher death rates, and experience more problems [9]. These health issues frequently lead to more intense care and longer hospital stays, which significantly raises the expense of healthcare. AMR also puts a burden on healthcare resources, requiring stricter infection control protocols as well as the use of more costly or hazardous alternative therapies that may be less successful and involve more hazards. Beyond just the direct cost of care, there are indirect costs as well. Increased sick days and long-term incapacity lower productivity [10].

4.5 The Role of Artificial Intelligence in Precision Medicine

The healthcare industry is adapting because of the efficiency and personalization that Artificial Intelligence (AI) provides to precision medicine [11]. Through AI's powerful data analysis skills, healthcare professionals can customize therapies to each patient's unique traits, including genetic profiles, environmental circumstances, and personal health histories. AI is being used in precision medicine for a variety of purposes, including using predictive models to forecast treatment outcomes and disease risks or evaluating genetic data to optimize pharmacological therapy [12]. These technologies optimize treatment regimens and minimize needless interventions, which not only improves treatment efficacy and diagnostic accuracy but also dramatically lowers healthcare expenditures. AI is becoming more and more important in precision medicine as it develops, offering more individualized, accurate, and efficient medical treatments. Redefining therapeutic approaches and patient care standards is expected to result from the convergence of AI and precision medicine, which will ultimately improve patient outcomes and streamline healthcare systems. Artificial Intelligence (AI) has rapidly permeated many facets of healthcare, transforming how medical data is interpreted, diagnoses are made, and treatments are carried out. AI is currently employed in healthcare in a variety of ways, including drug development, genomics, predictive analytics, and medical imaging [13]. Big amounts of data can be processed by AI algorithms at rates and precision faster than by humans (Fig. 2).

Machine learning models, for example, have proven to be remarkably effective at diagnosing diseases from imaging data. They can identify malignancies in radiology scans with a precision that is on par with, or in some cases even better than, that of expert radiologists [14]. AI technologies have a great deal to offer precision medicine, which focuses on customizing medical care to each patient's unique needs. Artificial intelligence (AI) may examine genetic data, lifestyle data, and clinical histories to find trends that would not be noticeable to human observers [15]. AI is quite useful in predicting illness susceptibility and even in tailoring treatment strategies because of its capacity. Based on each patient's unique genetic profile, AI models assist in determining which patients are more likely to respond to therapies, which may result in less invasive and more effective medical interventions [16]. Furthermore, these prediction models can be dynamically improved by AI, which can learn from fresh data and improve over time, increasing the accuracy of the models.

Personalized treatment plans, improved outcomes for patients, and increased diagnostic accuracy are just a few advantages of using AI in therapeutic approaches. Artificial intelligence (AI) systems are capable of rapidly analyzing intricate biomedical data and making personalized therapy recommendations that maximize efficacy and reduce negative effects [17]. For instance, in oncology, AI algorithms can forecast which chemotherapy medications will work best for a certain patient's tumor type by considering genetic alterations and past treatment outcomes. AI is also capable of real-time monitoring and management of chronic illnesses, modifying treatment plans as necessary to avoid problems and enhance quality of life [18].

4.6 Precision Treatment Strategies

Particularly in the context of the growing prevalence of antimicrobial resistance (AMR), precision treatment approaches in antimicrobial therapy constitute a substantial breakthrough in the fight against

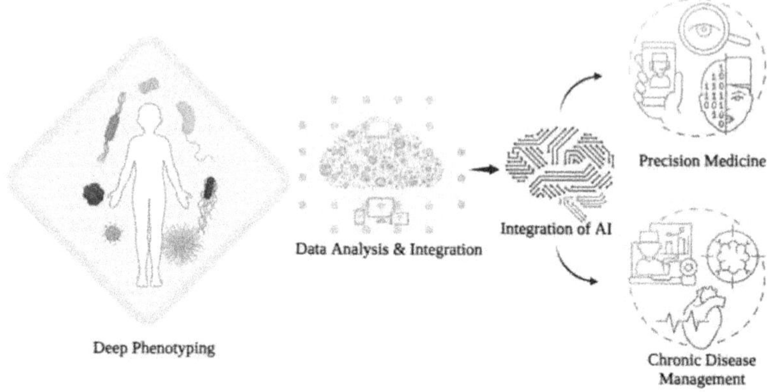

Figure 2: Workflow of AI in precision medicine. It illustrates the workflow of integrating Artificial Intelligence (AI) into precision medicine. On the left, it depicts the deep Phenotyping. which involves the detailed characterization of an individual's biological and physiological traits. This data is then fed into a process labeled data analysis & integration, depicted as a cloud filled with interconnected data symbols. The analyzed data is further processed with AI, shown as a stylized brain symbol representing the integration of AI. The AI integration leads to two outcomes: precision medicine, and chronic disease management, indicating the application of AI for personalized treatment and long-term disease management.

infections. This strategy makes use of artificial intelligence (AI) to find new therapeutic targets, optimize antibiotic dosage schedules, and customize treatments to the unique characteristics of each patient. By using artificial intelligence (AI) to evaluate a patient's genetic information, previous treatment outcomes, and other clinical variables, personalized antimicrobial therapy can choose the best medications, improving treatment outcomes and lowering the likelihood of resistance. AI-driven models also assist in adjusting the dosage of antibiotics to attain the best possible therapeutic levels while reducing side effects and avoiding misuse. Furthermore, AI's capacity to sort through enormous datasets speeds up the discovery of new therapeutic targets and may help develop new medicines that are more potent against types of bacteria that have developed resistance. When combined, these precision approaches represent a revolutionary change in the way infections are managed, providing more individualized, efficient, and long-lasting treatment choices. This section explores the latest developments and uses of AI in antimicrobial therapy, highlighting the critical role that AI will play in determining treatment paradigms in the future.

4.6.1 Personalized Antimicrobial Therapy using AI

Personalized antimicrobial therapy is a novel strategy that customizes treatment to meet the specific needs of each patient, made possible by the introduction of artificial intelligence (AI) in the healthcare industry. To determine which antimicrobial drugs are best for each patient, AI systems examine enormous databases that include genomic information, previous treatment responses, and present clinical results. Personalized medicine reduces antibiotic usage by ensuring the proper medication is given to the right patient at the right time, which helps fight antimicrobial resistance (AMR) [19]. Recent work highlights how AI can predict pathogen-specific medication susceptibility, improve treatment outcomes, and reduce the emergence of resistance. For instance, image-based techniques can be used to identify bacteria that are resistant to antibiotics, while data-driven methods can be used to anticipate new antibiotic compounds. More efficient ways to find more promising antimicrobial drugs include using AI-assisted compound library screening or designing novel compound structures. AI may also be used to anticipate probable resistance locations and associated enzyme functions using available data, such as genetic data, which paves the way for developing more effective antibiotics. AI has also made target discovery and dynamic modeling, peptide synthesis and design, SAR and STR evaluation, and medication repurposing easier. In the ensuing sections, we will organize antimicrobials into four main categories: phage therapy, AMPs, small compounds, and essential oils (EOs) [20].

4.6.2 Optimization of Antibiotic Dosing Regimens

AI is also essential for optimizing the dosage of antibiotics. Conventional dosage methods sometimes take a standardized approach, which can result in either an overdose or an underdose, both of which are factors in the emergence of AMR. The ideal dosage can be predicted by machine learning algorithms using patient-specific parameters including body weight, age, renal function, and genetic information. These models aid in optimizing the drug's effectiveness while reducing adverse effects and the possibility of resistance, as well as obtaining the intended pharmacokinetic and pharmacodynamic results. The achievement of these ideal dose regimens in clinical settings using AI-driven models is demonstrated in a previous study, which improves patient outcomes and makes more efficient use of antibiotics. While deep learning (DL) algorithms learn feature representations from raw input data, traditional AI and machine learning (ML) techniques rely on explicitly created features [21]. For drug response prediction, DL-based models perform better than conventional ML-based models; nevertheless, their generalizability and interpretability still need to be improved. This competing purpose can be addressed with the aid of multi-objective optimization techniques, network training, and proper representation design. Another issue with DL models with numerous layers is their interpretability; interpretable DL model strategies should be included in future research. Drug Response P rediction (DRP) is a domain where DL approaches often beat ML techniques. DL networks provide a range of designs that are currently being investigated [19,20].

4.6.3 Identification of Novel drug Targets through AI-driven Research

The development of new antibiotics is dependent on the identification of novel therapeutic targets, which is made possible by AI-driven research as drug resistance to current treatments keeps increasing. Artificial Intelligence systems can examine biological and chemical data to identify previously unknown or unfeasible potential targets. By analyzing enormous databases of bacterial protein interactions and structures, for instance, deep learning techniques have been utilized to find novel bacterial weaknesses that could be the focus of future antimicrobial treatments. This approach improves the accuracy and specificity of the medications created while also quickening the drug discovery process. A recent study claims a breakthrough in this field, using AI to uncover a novel mode of action against bacterial strains that are resistant to many drugs, providing hope in the fight against superbugs [22].

Machine learning potentials that can perform high-throughput calculations with DFT accuracy or better in millisecond time scales will speed up the search for ligand conformations and help prevent false positives and false negatives. As bulk phase protein/ligand simulation advances, it will enable precise QM free energy of binding computations that require little human involvement during parameterization. For high-throughput screening applications, *de novo* molecular design may be able to provide precise predictions of lead compounds to target for simulation, hence reducing the search space. Better, higher yield and more affordable synthetic pathways are anticipated with automated synthesis planning. A fully automated drug discovery pipeline powered by AI might someday be possible with the merging of the techniques covered in this view with other techniques that have not been covered, like robotic synthesis [17,23].

4.7 Overcoming Challenges and Limitations

Navigating a complicated terrain of technological, ethical, and regulatory difficulties is necessary to implement AI-driven precision medicine. Overcoming challenges about data quality, interoperability, ethical issues, and regulatory compliance is necessary for the effective integration of AI in healthcare. To fully utilize AI technology to improve patient care and treatment outcomes, these concerns must be addressed. A comprehensive overview of these issues is provided in this part, along with solutions and their significance for upcoming developments in precision medicine.

Ensuring ideal data quality and accessibility are the main obstacles to the implementation of AI in healthcare. For AI systems to train useful models, substantial amounts of precise and thorough data are needed. However, problems with incompleteness, correctness, and consistency are common in healthcare data. Additionally, data is often fragmented among many healthcare systems, making data exchange challenging and diminishing the efficacy of AI applications. It is imperative to make efforts to improve data-sharing methods and standardize data formats. To enhance the integration of data and facilitate the more efficient application of AI, for instance, programs like the Health Level Seven International (HL7) Fast Healthcare Interoperability Resources (FHIR) standard are being undertaken [24].

4.7.1 Ethical considerations in AI-driven precision medicine

Significant ethical issues, such as bias, privacy, and permission, are also brought up by the use of AI in precision medicine. Strict respect for ethical standards and privacy legislation is required when using personal health data in AI models [25]. Patients' consent must be obtained transparently, and they must be informed about how their data is used. Furthermore, there's a chance that AI systems will reinforce or magnify biases seen in training data, which could result in outcomes that are unfairly treated. Developing moral guidelines for the application of AI in healthcare and thoroughly evaluating AI systems for biases are necessary to address these concerns [18, 25].

4.7.2 Regulatory challenges and integration into clinical practice

A major obstacle to the use of AI in clinical practice is the presence of regulatory issues. Healthcare artificial intelligence regulations are still developing, with continuous discussions over responsibility, safety, and efficacy requirements. Regulators have to find a balance between patient safety and innovation, making sure that AI technologies are carefully inspected and assessed [26]. To fully integrate AI into clinical practice, healthcare professionals must be trained in the efficient use of AI tools and clinical workflows must be modified to accommodate AI-driven decision-making. To establish conditions that are favorable for the integration of AI in healthcare settings, cooperation between technologists, physicians, and regulators is crucial [12,26].

4.8 Future Directions and Emerging Trends

Several new developments and potential paths are influencing the course of AI-driven precision medicine as it develops. These include the development of AI algorithms for the management and prediction of antimicrobial resistance (AMR), the incorporation of genomic data into precision medicine, and a rise in international cooperation and joint ventures. In addition to promising to increase healthcare's effectiveness and efficiency, each of these fields also seeks to bring in a new era of global health advancements and individualized patient care. An in-depth discussion of these developments' possible effects and the advancements they offer in the area of precision medicine is provided in this section.

4.8.1 Advancements in AI algorithms for AMR prediction and management

Antimicrobial resistance (AMR), a rising worldwide health risk, is being predicted and managed by AI algorithms, which are becoming an increasing amount of efficient. Among these developments are machine learning models that, using phenotypic and genetic data, can anticipate the onset of resistance and make proactive modifications to treatment plans [27]. AI systems are being created, for instance, to examine patterns in the evolution of bacteria and forecast resistance pathways before they are useful for therapeutic use. These kinds of prediction abilities are essential for creating antimicrobial stewardship programs that work and for creating novel antimicrobials that are less prone to resistance. It is essential to conduct more research in this field to keep ahead of AMR trends and lessen their negative effects on public health [28].

4.8.2 Integration of genomic data into precision medicine approaches

The way diseases are identified and treated is being revolutionized by precision medicine's incorporation of genomic data. AI algorithms can now analyze enormous volumes of genomic data to detect illness susceptibilities and forecast individual responses to different therapies thanks to advancements in genomics and bioinformatics. Patient outcomes can be greatly improved by using this strategy, which enables more individualized treatment plans and more focused therapeutic choices. The goal of ongoing research is to improve the precision of genetic interpretations and broaden the situations for which they may be applied, from cancer to chronic illnesses like diabetes [29].

4.8.3 Collaborative efforts and international initiatives

International initiatives and strong collaborative efforts are necessary due to the global character of health concerns, especially when it comes to precision medicine and antimicrobial resistance. Global collaboration to standardize methods for genomic medicine and AI applications is being spearheaded by organizations like the World Health Organization (WHO) and the Global Genomic Medicine Collaborative (G2MC). These partnerships seek to standardize procedures, pool resources, and exchange expertise to more successfully address common concerns. Additionally, they prioritize closing gaps in healthcare access and making sure that everyone can benefit from AI-driven precision medicine, not only those in high-resource environments.

4.9 Case Studies

4.9.1 AMR Surveillance with Whole Genome Sequencing (WGS)

The use of Whole genome Sequencing (WGS) combined with AI has significantly advanced AMR surveillance. This strategy served as a paradigm for AMR surveillance and was prominently used during the COVID-19 pandemic to track and control the virus's transmission. With the help of the lessons learned from this pandemic, WGS is now widely used for more thorough infectious disease surveillance [30]. Real-time monitoring of pathogen evolution and medication resistance is made possible by the integration of WGS data with AI-enhanced analysis tools, which enables prompt and efficient public health actions. To improve surveillance capabilities across many industries and geographical areas, this approach highlights the necessity of international cooperation and data sharing while upholding the FAIR principles (findability, accessibility, interoperability, and reusability) [30].

4.9.2 Predicting AMR in E. Coli Using Machine Learning

Machine learning models were trained to predict antibiotic resistance in *E. Coli* in a multi-country study carried out in Africa [31]. SNP (Single Nucleotide Polymorphism) data taken from genomic files was analyzed using a variety of techniques, such as Random Forest, Gradient Boosting, and Logistic Regression. This method demonstrated how AI may improve diagnostic accuracy and direct treatment decisions in a variety of geographic contexts, highlighting the significance of efficient data encoding and model selection to properly forecast AMR [28]. Various methods and hyperparameter tuning were employed in complete machine learning procedures, which were crucial in maximizing the prediction models for real-world AMR surveillance applications.

References

[1] C. H. Rammelkamp, and T. Maxon T. Resistance of Staphylococcus aureus to the Action of Penicillin.∗. *https: //doi.org/103181/00379727-51-13986*. 1942; 51(3): 386–389. doi:10.3181/00379727-51-13986

[2] T. Pulingam, T. Parumasivam, A. Mohd Gazzali, *et al.* NC-ND license Antimicrobial resistance: Prevalence, economic burden, mechanisms of resistance and strategies to overcome. *Eur J Pharm Sci*. 2022; 170: 106103. doi:10.1016/j.ejps.2021.106103

[3] E. M. Darby, E. Trampari, P. Siasat, *et al.* nature reviews microbiology Molecular mechanisms of antibiotic resistance revisited. *Nat Rev Microbiol* |. 2023; 21: 280–295. doi:10.1038/s41579-022-00820-y

[4] M. A. Salam, M. Y. Al-Amin, and M. T. Salam, *et al.* Antimicrobial Resistance: A Growing Serious Threat for Global Public Health. *Healthc 2023, Vol 11, Page 1946*. 2023; 11(13): 1946. doi:10.3390/HEALTHCARE11131946

[5] R. J. Ludwig, A. A. Fayad, E. Anes, *et al.* Evolutionary Approaches to Combat Antibiotic Resistance: Opportunities and Challenges for Precision Medicine. *Front Immunol* | *www.frontiersin.org*. 2020; 1. doi:10.3389/fimmu.2020.01938

[6] T. F. Landers, B. Cohen, T. E.Wittum, and E. L. Larson. A review of antibiotic use in food animals: Perspective, policy, and potential. *Public Health Rep.* 2012; 127(1): 4–22. doi:10.1177/003335491212700103

[7] P. Martins Da Costa, L. Loureiro, A. J. F. Matos. Transfer of Multidrug-Resistant Bacteria Between Intermingled Ecological Niches: The Interface Between Humans, Animals and the Environment. *OPEN ACCESS Int J Environ Res Public Heal.* 2013; 10: 10. doi:10.3390/ijerph10010278

[8] H. Endale, M. Mathewos, and D. Abdeta. Potential Causes of Spread of Antimicrobial Resistance and Preventive Measures in One Health Perspective-A Review. *Infect Drug Resist.* 2023; 16: 7515–7545. doi:10.2147/IDR.S428837

[9] S. Kumar. Antimicrobial resistance: a top ten global public health threat. *EClinicalMedicine*. 2021; 41: 101221. doi:10.1016/j.eclinm.2021.101221

[10] S. Khalid Ahmed, S. Hussein, K. Qurbani, *et al.* Antimicrobial resistance: Impacts, challenges, and future prospects. *J Med Surgery, Public Heal.* 2024; 2: 100081. doi:10.1016/j.glmedi.2024.100081

[11] K. B. Johnson, W. Q. Wei, D. Weeraratne, *et al.* Precision Medicine, AI, and the Future of Personalized Health Care. *Clin Transl Sci.* 2021; 14(1): 86-93. doi:10.1111/CTS.12884

[12] B. Derraz, G. Breda, C. Kaempf, *et al.* New regulatory thinking is needed for AI-based personalised drug and cell therapies in precision oncology. doi:10.1038/s41698-024-00517-w

[13] F. Boniolo, E. Dorigatti, A. J. Ohnmacht, D. Saur, B. Schubert, M.P. Menden. Artificial intelligence in early drug discovery enabling precision medicine. *Expert Opin Drug Discov.* 2021; 16(9): 991–1007. doi:10.1080/17460441.2021.1918096

[14] T. Obafemi-Ajayi, A. Perkins, B. Nanduri, *et al.* No-boundary thinking: a viable solution to ethical data-driven AI in precision medicine. *AI Ethics 2021 24.* 2021; 2(4): 635–643. doi:10.1007/S43681-021-00118-4

[15] F. S. Collins, and H. Varmus. A new initiative on precision medicine. *N Engl J Med.* 2015; 372(9): 793–795. doi:10.1056/NEJMP1500523

[16] A. Blasiak, J. Khong, and T. Kee. CURATE.AI: Optimizing Personalized Medicine with Artificial Intelligence. *SLAS Technol.* 2020; 25(2): 95–105. doi:10.1177/2472630319890316/ASSET/IMAGES/LARGE/10.1177_2472630319890316-FIG2.JPEG

[17] N. J. Schork. Artificial Intelligence and Personalized Medicine. *Cancer Treat Res.* 2019; 178: 265-283. doi:10.1007/978-3-030-16391-4_11

[18] N. Naik, B. M. Z. Hameed, D. K. Shetty, *et al.* Legal and Ethical Consideration in Artificial Intelligence in Healthcare: Who Takes Responsibility? *Front Surg.* 2022; 9. doi:10.3389/fsurg.2022.862322

[19] M. S. M. Serafim, T. Kronenberger, P. R. Oliveira, *et al.* The application of machine learning techniques to innovative antibacterial discovery and development. *Expert Opin Drug Discov.* 2020; 15(10): 1165–1180. doi:10.1080/17460441.2020.1776696

[20] R. Shouval, O. Bondi, H. Mishan, A. Shimoni, R. Unger, A. Nagler. Application of machine learning algorithms for clinical predictive modeling: a data-mining approach in SCT. *Bone Marrow Transplant.* 2014; 49: 332–337. doi:10.1038/bmt.2013.146

[21] S. J. Maceachern, N. D. Forkert. Machine learning for precision medicine. *Genome*. 2021; 64(4): 416–425. doi:10.1139/GEN-2020-0131/ASSET/IMAGES/LARGE/GEN-2020-0131F1.JPEG

[22] M. C. R. Melo, J. R. M. A. Maasch, C. de la Fuente-Nunez. Accelerating antibiotic discovery through artificial intelligence. *Commun Biol 2021 41*. 2021; 4(1): 1–13. doi:10.1038/s42003-021-02586-0

[23] A. Talat, and A. U. Khan. Artificial intelligence as a smart approach to develop antimicrobial drug molecules: A paradigm to combat drug-resistant infections. *Drug Discov Today*. 2023; 28(4): 103491. doi:10.1016/J.DRUDIS.2023.103491

[24] S. Maxhelaku, and A. Kika. IMPROVING INTEROPERABILITY IN HEALTHCARE USING HL7 FHIR. Published online August 22, 2019. doi:10.20472/IAC.2019.047.012

[25] D. S. Char, N.H. Shah, and D. Magnus. Implementing Machine Learning in Health Care—Addressing Ethical Challenges. *N Engl J Med*. 2018; 378(11): 981–983. doi:10.1056/NEJMP1714229/SUPPL_FILE/NEJMP1714229_DISCLOSURES.PDF

[26] F. Jiang, Y. Jiang, H. Zhi, *et al.* Artificial intelligence in healthcare: past, present and future. *Stroke Vasc Neurol*. 2017; 2(4): 230–243. doi:10.1136/SVN-2017-000101

[27] A. Howard, S. Aston, A. Gerada, *et al.* Antimicrobial learning systems: an implementation blueprint for artificial intelligence to tackle antimicrobial resistance. *Lancet Digit Heal*. 2024; 6(1): e79–e86. doi:10.1016/S2589-7500(23)00221-2

[28] T. Ali, S. Ahmed, and M. Aslam. Artificial Intelligence for Antimicrobial Resistance Prediction: Challenges and Opportunities towards Practical Implementation. *Antibiotics*. 2023; 12(3). doi:10.3390/antibiotics12030523

[29] K. S. Lau-Min, D. McKenna, S. B., Asher, *et al.* Impact of integrating genomic data into the electronic health record on genetics care delivery. *Genet Med*. 2022; 24(11): 2338–2350. doi:10.1016/J.GIM.2022.08.009

[30] M. J. Struelens, and C. Ludden. Real-time genomic surveillance for enhanced control of infectious diseases and antimicrobial resistance. 2024; (April): 1–23. doi:10.3389/fsci.2024.1298248

[31] M. Nsubuga, R. Galiwango, D. Jjingo, and G. Mboowa. Generalizability of machine learning in predicting antimicrobial resistance in E. coli: a multi-country case study in Africa. *BMC Genomics*. 2024; 25(1): 1–13. doi:10.1186/S12864-024-10214-4/TABLES/5

5

Artificial Intelligence-driven Diagnostics for Antimicrobial Resistance

Himanshu,[1,2] *Ramendra Pati Pandey*[4] and *Chung-Ming Chang*[2,3,*]

The growing threat of antimicrobial resistance (AMR) demands innovative diagnostic approaches. Traditional diagnostic methods often fall short in providing timely and accurate results, necessitating more advanced solutions. This chapter explores the pivotal role of artificial intelligence in revolutionizing AMR diagnostics. AI techniques, particularly machine learning and deep learning are emphasized for their superior precision and speed in AMR detection. This chapter discusses various ML algorithms, such as supervised, unsupervised, and reinforcement learning, and their specific applications in identifying AMR. Despite the advancements, challenges in data acquisition and

[1] Graduate Institute of Biomedical Sciences, Chang Gung University, No. 259, Wenhua 1st Road, Guishan District, Taoyuan city 33302, Taiwan (R.O.C).

[2] Master & Ph.D. Program in Biotechnology Industry, Chang Gung University, No. 259, Wenhua 1st Road, Guishan District, Taoyuan city 33302, Taiwan (R.O.C).

[3] Department of Medical Biotechnology and Laboratory Science, Chang Gung University, No. 259, Wenhua 1st Road, Guishan District, Taoyuan city 33302, Taiwan (R.O.C).

[4] School of Health Sciences and Technology (SoHST), UPES, Bidholi, Dehradun 248007, Uttarakhand, India.

* Corresponding author: cmchang@mail.cgu.edu.tw

algorithm optimization persist, underscoring the need for ongoing research and validation. The chapter concludes by advocating for the integration of AI-driven diagnostics into clinical workflows and calling for increased research funding to effectively combat the AMR crisis. These advanced methods can process vast amounts of data quickly and accurately, identifying patterns and predicting resistance with greater efficiency.

5.1 Introduction

Antimicrobial Resistance (AMR) occurs naturally when microorganisms such as bacteria, viruses, fungi, and parasites gradually become capable of resisting the effects of drugs designed to eliminate them from the body. In the process of promoting the development of AMR, antibiotics have been drastically and wrongly used in a wide range of areas inclusive of human health care, animal husbandry and environment. This subsequently renders erstwhile effective therapies impotent causing prolongation of diseases, higher numbers of deaths and increased economic burdens on patients and their families [1]. These circumstances have compelled authorities to implement essential measures to address these challenges and advocate for suitable diagnostic methodologies. These strategies aim to facilitate the adoption of effective measures to mitigate the risk of AMR, incorporating preventive measures and judicious antibiotic use. Yet, identifying antibiotic resistant genes presents challenges, demanding both high accuracy for effective treatment and robustness for swift problem identification [2]. Presently, laboratory-based characterization and diagnostic techniques lack the depth required for comprehensive surveillance, leading to insufficient information. Moreover, tests yield inconsistent results influenced by environmental factors and laboratory conditions [3]. Combining data from high-throughput sequencing, such as whole-genome sequencing, with laboratory techniques offers a promising avenue for exploring genetic variations associated with prevalent AMR. An essential element in combating AMR is identifying infectious pathogens early, distinguishing between infectious and non-infectious pathologies, and effectively managing any complications that arise.

In recent times, technological advancements have led to a significant enhancement in the computing power and data storage capacity of computers, enabling the processing and analysis of vast amounts of data within short time frames. AMR-AI models offer the potential to diminish uncertainty by rapidly predicting and addressing AMR compared to the predominantly slow culture-based methods utilized in microbiology laboratory diagnostics. These predictive capabilities have the capacity to enhance clinical decision support through AMR-AI models, potentially enabling more precise targeting of antimicrobial therapy and implementation of protective isolation measures. This proactive approach aims to mitigate the risks associated with uncontrolled AMR infections, unnecessary selection pressures on AMR, and subsequent transmission of resistant pathogens [4]. Consequently, A rtificial I ntelligence (AI) techniques, particularly M achine L earning (ML) and D eep L earning (DL), are being increasingly utilized across various domains [5,6]. The medical sector has witnessed a surge in the adoption of AI, with the development of the metaverse for intelligent healthcare highlighted [7].

5.2 Traditional Diagnostic Methods for AMR and their Limitations

Traditional diagnostics methods for AMR rely on culturing pathogens from clinical samples and performing susceptibility testing to determine their response to antimicrobial agents. These methods have been the cornerstone of clinical microbiology for decades and include several techniques such as: Phenotypic testing, molecular, and spectroscopic methods.

5.2.1 Phenotypic testing

Phenotypic methods for detecting AMR entail directly observing the impact of antimicrobial drugs on bacterial growth within a laboratory environment. These methods enable the determination of the minimum inhibitory concentration (MIC) of a specific drug for a particular bacterial strain, representing the lowest concentration of the drug capable of inhibiting bacterial growth. Various phenotypic methods are available for AMR detection, including broth, macro broth or tube-dilution methods, agar dilution, and disk diffusion.

Phenotypic methods are widely regarded as reliable and accurate; nonetheless, they can be time-consuming and labor-intensive [8]. Moreover, they necessitate the use of live bacteria, which could raise safety concerns. Additionally, these methods may not be suitable for testing certain bacterial species with distinct growth requirements or specific resistance mechanisms that are not readily observable through standard growth-based assays.

5.2.2 Molecular diagnostics

Molecular-based methods for AMR detection entail identifying specific genetic markers linked to resistance against antimicrobial drugs. These markers may reside within the bacterial genome or on plasmids, which are small, circular fragments of DNA capable of being transferred between bacteria. This encompasses techniques such as Polymerase Chain Reaction (PCR), DNA microarray technology, which employs a DNA chip or microarray to detect the presence of numerous genes or mutations simultaneously in a single sample, and Whole Genome Sequencing (WGS).

Molecular-based methods offer several advantages over phenotypic-based methods for AMR detection. They are typically faster and more sensitive, capable of detecting resistance to a wide array of antimicrobial drugs. Moreover, these methods furnish detailed insights into the genetic underpinnings of resistance and can aid in tracking the dissemination of resistant strains within a population. However, molecular-based methods may necessitate specialized equipment and expertise and may not be suitable for all bacterial types [9]. Additionally, they generally incur higher costs than phenotypic-based methods and may not be feasible for large-scale screening programs.

5.2.3 Spectroscopic methods

Spectroscopy techniques are utilized in the detection and characterization of AMR in microorganisms within spectroscopic-based methods. These methods provide rapid and high-throughput results, facilitating the identification and quantification of resistance genes and proteins. In the following section, we will delve into the most employed spectroscopic-based techniques for detecting and characterizing AMR in microorganisms. These methods encompass Matrix-assisted laser desorption/ionization time-of-flight (MALDI-TOF), Fourier-transform infrared spectroscopy (FTIR), Nuclear Magnetic Resonance (NMR) spectroscopy, Raman Spectroscopy, and Surface-Enhanced Raman Spectroscopy.

Spectroscopic techniques offer valuable insights into AMR detection, yet they face limitations. Sensitivity challenges arise with low-level AMR detection, hindering early-stage diagnosis. Specificity issues may blur distinctions between resistance mechanisms or strains. Labor-intensive sample preparation risks introducing artifacts. Complex data necessitates expertise in both spectroscopy and microbiology for accurate interpretation. High costs for equipment and maintenance limit accessibility, especially in resource-limited settings. Spectroscopy's inability to cover the full spectrum of resistance mechanisms poses a challenge, as does susceptibility to sample matrix effects. Integrating spectroscopic methods into clinical workflows requires extensive validation and regulatory approval, slowing adoption despite their potential impact on AMR diagnostics.

5.3 AI Techniques for AMR Detection

AI, with its capacity to process vast amounts of data and convert it into actionable insights and knowledge, surpasses human cognitive abilities, addressing the limitations of rational decision-making caused by incomplete information and time constraints. AI, when appropriately developed, operates free from behavioral constraints, including irrational deviations from guidelines, peer influence, hierarchical cultural norms, and fatigue. Algorithms can learn objectively and often yield more accurate predictions than those derived from human practice alone.

5.4 Machine Learning Algorithms for AMR Prediction

Drug resistance is a complex challenge of the 21st century, arising from the intricate interplay of evolutionary forces, microbiology, the built environment, and human behavior. Machine learning ML, a subfield of AI, has emerged as a potential approach to tackle this complexity. ML involves the development of algorithms capable of building predictive models using training data sets with minimal human intervention. These algorithms are broadly categorized into supervised ML, where they train on labeled data, unsupervised ML, where they learn structures within the data set independently, and reinforcement learning, where they optimize for desired outcomes as new data are integrated continuously. Compared to traditional statistical modeling, ML algorithms prioritize prediction on unseen data rather than comprehending data-generating processes. However, recent advancements have also led to the development of ML algorithms for causal inference with observational data [10,11].

In 1959, Arthur Samuel described ML as "the field of study that gives computers the ability to learn without being explicitly programmed" [12]. Later, in 1997, Tom Mitchell provided a more engineering-oriented definition, stating that a "computer program is said to learn from experience (E) with respect to some task (T) and some performance measure (P) if its performance on (T), as measured by (P), improves with experience (E)" [13]. Recent advancements in digitizing medical records and experimental data have facilitated the integration of ML in biology and medicine. Clinical trials are increasingly adopting ML processes to enhance the efficiency and quality of clinical research and pre clinical drug development [14]. ML techniques are also applied to assess the risk of developing sepsis based on patients' clinical records [15]. At the cellular level, convolutional neural networks (CNNs) are utilized to predict interactions between transcription factors and histones within chromosome structures, aiding in the analysis of genome architecture and gene regulation [16]. Additionally, neural networks are employed to identify the role of non-coding DNA in human gene expression regulation [17], while recurrent neural networks (RNNs) are utilized to characterize chromatin folding in *Drosophila melanogaster* [18]. Moreover, the availability of large-scale high-throughput genomic and epigenomic data has led to numerous studies showcasing the potential applications of ML in genomics [19] and non-coding RNAs [20]. ML is also utilized to assist clinicians in treating infectious diseases [21]. However, the use of ML in studying drug-resistant pathogens is still in its early stages (refer to Fig. 1).

ML can indeed be categorized into four main types: supervised learning, unsupervised learning, semi-supervised learning, and reinforcement learning. These will be discussed in following sections:

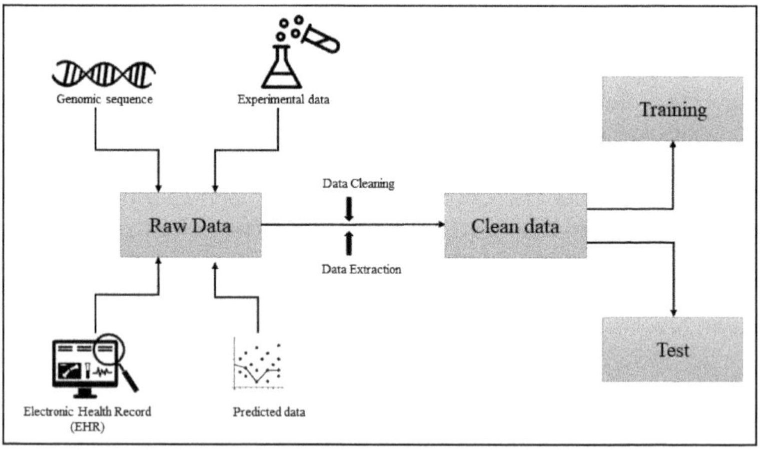

Figure 1: ML Analyses in Detection of Antimicrobial Resistance.

5.4.1 Supervised learning

The principle of supervised ML plays an essential role in developing accurate prediction models for bacteria identification. This algorithm operates by leveraging a known dataset of characteristics along with labeled microbial strain categories as a training set. During the training process, the algorithm establishes relationships between the characteristics and the microbial categories, ultimately constructing a relevant mathematical model. Supervised learning techniques can be broadly categorized into two main types: "classification" and "regression". [22,23].

In classification problems, the objective is to categorize future inputs into predefined classes by training on labeled examples, where each input is associated with its corresponding category [24]. With sufficient high-quality training data, models generated using classification techniques can accurately classify future data, without the need for explicit programming of input data details [25,26]. For example, researchers may seek to automate the identification of cell species in microscopy images without human intervention. By using a training dataset consisting of microscopy images of various species labeled with their names, a classification model can learn the relationships between visual features of species and their labels. Subsequently, the trained model can be applied to unlabeled microscopy images to identify species, saving researchers time and effort while facilitating the sharing of the model within the scientific community. Classification learning algorithms are versatile and applicable to diverse types of data, not limited to images; any data that can be categorized into predefined classes can be utilized for training a classification model.

Classification methods are designed to predict discrete class labels for inputs, whereas regression methods aim to predict continuous numerical values for given numerical inputs. In regression techniques, the model learns from training data consisting of numerical inputs and their corresponding numerical outputs. Consequently, the resulting model establishes a continuous mathematical relationship between inputs (independent variables) and outputs (dependent variables) [26]. This model can then be utilized to make numerical predictions for future inputs. By training a regression model on a comprehensive dataset of input-output pairs, the researcher can develop a model capable of accurately predicting numerical outputs for future inputs. This not only offers practical utility but also aids fundamental research by providing precise mathematical and physical relationships that can be further analyzed and understood in terms of theoretical concepts [27] (Fig. 2).

5.4.2 Unsupervised learning

Unsupervised learning methods differ from supervised learning by attempting to learn patterns and structures from unlabeled data [28]. This primarily involves techniques such as data clustering, although other methods like anomaly detection and dimensionality reduction also fall under this category. Clustering algorithms aim to partition unlabeled data into groups with similar characteristics, aiding in the extraction of valuable insights from high-dimensional data that may be challenging for humans to process. Anomaly detection, on the other hand, focuses on identifying outliers in large datasets by comparing data points to learned patterns, particularly useful for handling noisy experimental data [29]. Dimensionality reduction methods seek to simplify high-dimensional data while retaining essential information, facilitating easier analysis and utilization of such data (Fig. 2).

Unsupervised learning techniques can also be complemented by supervised learning in a hybrid approach known as "semi-supervised" learning, which leverages partially labeled data [30]. This becomes particularly advantageous when dealing with extensive datasets where manually labeling every data point is impractical. Various examples of unsupervised learning methods include k-means clustering [31], hierarchical clustering [32], DBSCAN [33], isolation forests [34], principal component analysis [35], auto encoders [36], etc.

5.4.3 Reinforcement learning

Reinforcement learning relies on the concept of learning from 'rewards' acquired through interactions with an environment. These problems are typically formulated as discrete-time stochastic control processes known as 'Markov decision processes', aiming to train a computational system (or 'agent') to determine the optimal strategy (or 'policy') for achieving a defined objective. The environment is characterized by 'states' that the agent can occupy, while the agent can execute specific 'actions' to interact with the environment [37]. During these interactions, the agent collects numerical rewards that quantify its performance for executing particular actions. The objective of the agent is to maximize these rewards (utilizing advanced statistical methods) by learning the

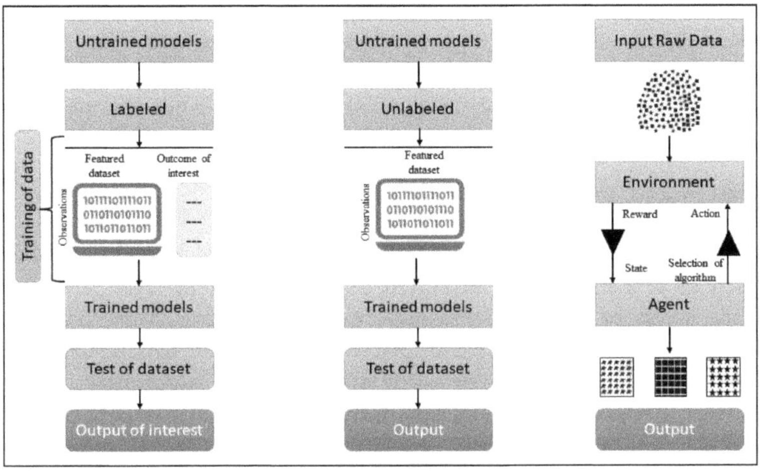

Figure 2: Categories of ML Analyses.

most effective policy for decision-making in various scenarios through iterative interactions with its environment [38].

AlphaZero and AlphaGo [39] represent examples of model-based approaches in reinforcement learning. Conversely, model-free approaches do not rely on the distribution of transition probabilities, or the reward function associated with Markov Decision Processes (MDP). Algorithms like Q-learning, Deep Q Network, Monte Carlo Control, and SARSA (State–Action–Reward–State–Action) are typical instances of model-free algorithms [40]. A critical distinction between model-free and model-based learning lies in the presence of a policy network, which is required for model-based RL but not for model-free approaches (Fig. 2).

Popular Machine Learning (ML) Algorithms in the Field of AMR

Supervised machine learning algorithms commonly utilized include linear regression, logistic regression, k-nearest neighbors (k-NN), support vector machine (SVM), decision tree (DT), random forest (RF), and gradient boosting machine (GBM). These algorithms are among the most prevalent, followed by neural networks and deep learning approaches. The primary performance metric used in ML-driven prediction models is the area under the receiver operating characteristic curve (AUROC).

Additionally, other performance metrics such as accuracy, sensitivity, specificity, positive and negative predictive values, precision, recall, and the F1 measure are employed. These metrics can be derived from a confusion matrix, a tabular structure summarizing the hits and misses of a learning algorithm [41].

5.5 Deep Learning

DL is a subset of ML that employs multi-layered neural networks. Each layer's output is transmitted to the subsequent layer, allowing these layers to extract increasingly complex features from the input data [42]. DL models excel in discovering intricate patterns in data, which may pose challenges for conventional ML algorithms, owing to their hierarchical learning approach. A key distinction between DL and ML lies in the degree of abstraction they achieve. ML algorithms often utilize low-level aspects of data, such as individual words in text documents or pixel values in images. In contrast, DL models can learn abstract representations of data, capturing high-level concepts like object categories or semantic meanings. Additionally, the amount of data required for training models differs between ML and DL. DL models typically necessitate large datasets to learn complex patterns, whereas ML algorithms can be effective with smaller datasets [43].

DL excels in analyzing vast and diverse datasets, encompassing speech, language, and images. Its application in healthcare and medical science has yielded numerous beneficial health outcomes [44]. These sophisticated computational models possess the capability to discern and extract features from extensive and varied datasets. Noteworthy DL algorithms include Deep Neural Networks (DNN), Convolutional Neural Networks (CNN), You Only Look Once (YOLO), among others [45]. Recently, researchers have leveraged various neural network models such as CNN, Long short-term memory (LSTM), CNN-LSTM, and Bidirectional LSTM for genome and sequence-associated research endeavors. Gunasekaran et al., for instance, employed these algorithms to classify DNA sequences, subsequently evaluating them using diverse classification metrics. Their findings underscored the high accuracy rates achieved by these DL models [46].

Contemporary genomics studies and molecular diagnostics have enabled the identification of antibiotic-resistant genes from microbial genomes, which can serve as valuable training data for DL-based diagnostic methods. Various antimicrobial resistance genes have been discovered in antibiotic-resistant microbes found in hospital settings. By leveraging genome sequencing of antibiotic-resistant microbes, these resistance genes can be incorporated into training datasets for DL models. For instance, Zhang et al. recently developed a DL model capable of predicting Next-Generation Sequencing (NGS) outcomes using DNA probe sequences, demonstrating its effectiveness in handling extensive datasets [47]. These DL models facilitate rapid diagnosis of antibiotic resistance and can predict antimicrobial susceptibility from patient samples, aiding in the development of prompt and targeted treatments for emergency cases of microbial infections. Ultimately, DL models contribute to improved understanding and management of Multiple Drug Resistance (MDR) infections among emergency patients in hospital settings (refer to Fig. 3).

While DL models have seen limited applications in addressing AMR until recently, challenges remain in collecting clinical data on antibiotic resistance-associated microbes for genome sequencing and database development. Establishing training datasets for DL models requires meticulous data collection in standardized research settings. We call upon policymakers worldwide to allocate research funding for advanced DL research aimed at addressing the global antibiotic resistance crisis. Additionally, we encourage health professionals and scientists to devise effective strategies and conduct focused DL research to combat antibiotic resistance in treatment-related emergencies associated with microbial infections [48].

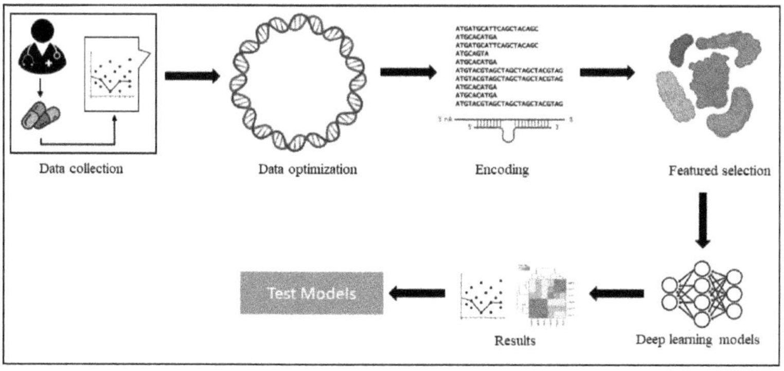

Figure 3: Deep learning to detect AMR.

5.6 Challenges and Future Directions

This chapter introduces AI diagnostics in the realm of AMR, highlighting numerous research issues within the field. Consequently, in this section, we aim to outline and deliberate on the challenges encountered, as well as explore potential research opportunities and future directions.

In general, the effectiveness and efficiency of a machine learning-based solution relies heavily on the nature and characteristics of the data, as well as the performance of the learning algorithms. However, collecting relevant data in domains such as cybersecurity, IoT, healthcare, and agriculture can be challenging. Despite the abundance of data available in the current cyberspace, obtaining useful data for specific machine learning-based applications, such as smart city applications, and managing it effectively is not straightforward. Therefore, a more comprehensive examination of data collection methods is essential when working with real-world data.

To analyze data and extract insights, a multitude of machine learning algorithms are available. However, selecting the appropriate learning algorithm for a specific application can be challenging. This difficulty arises because the outcomes of different learning algorithms may vary depending on the characteristics of the data. Choosing an inappropriate learning algorithm can result in unexpected outcomes, potentially leading to wasted effort and diminished model effectiveness and accuracy. Therefore, careful consideration and evaluation are necessary when selecting a learning algorithm for a given task.

Hence, the ultimate success of a machine learning-based solution and its corresponding applications hinges primarily on both the quality of the data and the efficacy of the learning algorithms. When the data is inadequate for learning—such as being non-representative, of poor quality, containing irrelevant features, or insufficient in quantity for training—the resulting machine learning or deep learning models may prove ineffective or yield lower accuracy. Therefore, proficiently processing the data and effectively leveraging a diverse set of learning algorithms are crucial for the development of machine learning-based solutions and the eventual creation of intelligent applications.

5.7 Conclusion

In conclusion, investing in AI-driven diagnostics is paramount for addressing the global challenge of AMR. The development of reliable, sensitive, and cost-effective diagnostic tools is essential in this endeavor.

Rapid diagnostic technologies, especially in primary care settings like rapid diagnostic tests, hold promise for improving targeted treatment strategies. Integrating advanced monitoring systems such as mobile applications with surveillance programs is crucial for effectively tracking AMR. Moreover, the incorporation of emerging approaches like machine learning and data mining, alongside automation, will play a pivotal role in advancing next-generation diagnostics. Continued research and investment in these areas are vital to effectively combating the threat of AMR and safeguarding global health.

References

[1] G. Y. Liu *et al.*, "Antimicrobial resistance crisis: could artificial intelligence be the solution?," *Military Medical Research 2024 11: 1*, vol. 11, no. 1, pp. 1–23, Jan. 2024, doi:10.1186/S40779-024-00510-1.

[2] A. A. Rabaan *et al.*, "Application of Artificial Intelligence in Combating High Antimicrobial Resistance Rates," *Antibiotics*, vol. 11, no. 6, p. 784, Jun. 2022, doi:10.3390/ANTIBIOTICS11060784.

[3] I. Díaz, "Machine learning in the estimation of causal effects: targeted minimum loss-based estimation and double/debiased machine learning," *Biostatistics*, vol. 21, no. 2, pp. 353–358, Apr. 2020, doi:10.1093/BIOSTATISTICS/KXZ042.

[4] E. H. Shortliffe and M. J. Sepúlveda, "Clinical Decision Support in the Era of Artificial Intelligence," *JAMA*, vol. 320, no. 21, pp. 2199–2200, Dec. 2018, doi:10.1001/JAMA.2018.17163.

[5] R. I. Mukhamediev, A. Symagulov, Y. Kuchin, K. Yakunin, and M. Yelis, "From Classical Machine Learning to Deep Neural Networks: A Simplified Scientometric Review," *Applied Sciences 2021, Vol. 11, Page 5541*, vol. 11, no. 12, p. 5541, Jun. 2021, doi:10.3390/APP11125541.

[6] C. Gupta *et al.*, "A Systematic Review on Machine Learning and Deep Learning Models for Electronic Information Security in Mobile Networks," *Sensors 2022, Vol. 22, Page 2017*, vol. 22, no. 5, p. 2017, Mar. 2022, doi:10.3390/S22052017.

[7] G. Joshi, A. Jain, S. R. Araveeti, S. Adhikari, H. Garg, and M. Bhandari, "FDA-Approved Artificial Intelligence and Machine Learning (AI/ML)-Enabled Medical Devices: An Updated Landscape," *Electronics (Switzerland)*, vol. 13, no. 3, p. 498, Feb. 2024, doi:10.3390/ELECTRONICS13030498/S1.

[8] M. Welker and A. van Belkum, "One System for All: Is Mass Spectrometry a Future Alternative for Conventional Antibiotic Susceptibility Testing?," *Front Microbiol*, vol. 10, p. 491958, Nov. 2019, doi:10.3389/FMICB.2019.02711/BIBTEX.

[9] R. Franco-Duarte *et al.*, "Advances in Chemical and Biological Methods to Identify Microorganisms—From Past to Present," *Microorganisms*, vol. 7, no. 5, p. 130, May 2019, doi:10.3390/MICROORGANISMS7050130.

[10] M. S. Schuler and S. Rose, "Targeted Maximum Likelihood Estimation for Causal Inference in Observational Studies," *Am J Epidemiol*, vol. 185, no. 1, pp. 65–73, Jan. 2017, doi:10.1093/AJE/KWW165.

[11] I. Díaz, "Machine learning in the estimation of causal effects: targeted minimum loss-based estimation and double/debiased machine learning," *Biostatistics*, vol. 21, no. 2, pp. 353–358, Apr. 2020, doi:10.1093/BIOSTATISTICS/KXZ042.
[12] A. L. Samuel, "Some Studies in Machine Learning Using the Game of Checkers," *IBM J Res Dev*, vol. 3, no. 3, pp. 210–229, Jul. 1959, doi:10.1147/RD.33.0210.
[13] M. Awad and R. Khanna, "Efficient learning machines: Theories, concepts, and applications for engineers and system designers," *Efficient Learning Machines: Theories, Concepts, and Applications for Engineers and System Designers*, pp. 1–248, Jan. 2015, doi:10.1007/978-1-4302-5990-9/COVER.
[14] E. H. Weissler *et al.*, "The role of machine learning in clinical research: transforming the future of evidence generation," *Trials*, vol. 22, no. 1, pp. 1–15, Dec. 2021, doi:10.1186/S13063-021-05489-X/FIGURES/3.
[15] D. R. Giacobbe *et al.*, "Validation of an Automated System for the Extraction of a Wide Dataset for Clinical Studies Aimed at Improving the Early Diagnosis of Candidemia," *Diagnostics*, vol. 13, no. 5, p. 961, Mar. 2023, doi:10.3390/DIAGNOSTICS13050961/S1.
[16] A. Jaroszewisz and J. Ernst, "An integrative approach for fine-mapping chromatin interactions," *Bioinformatics*, vol. 36, no. 6, pp. 1704–1711, Mar. 2020, doi:10.1093/BIOINFORMATICS/BTZ843.
[17] R. Movva, P. Greenside, G. K. Marinov, S. Nair, A. Shrikumar, and A. Kundaje, "Deciphering regulatory DNA sequences and noncoding genetic variants using neural network models of massively parallel reporter assays," *PLoS One*, vol. 14, no. 6, p. e0218073, Jun. 2019, doi:10.1371/JOURNAL.PONE.0218073.
[18] M. B. Rozenwald, A. A. Galitsyna, G. V. Sapunov, E. E. Khrameeva, and M. S. Gelfand, "A machine learning framework for the prediction of chromatin folding in Drosophila using epigenetic features," *PeerJ Comput Sci*, vol. 6, pp. 2–21, 2020, doi:10.7717/PEERJ-CS.307/SUPP-3.
[19] A. Talukder, C. Barham, X. Li, and H. Hu, "Interpretation of deep learning in genomics and epigenomics," *Brief Bioinform*, vol. 22, no. 3, pp. 1–16, May 2021, doi:10.1093/BIB/BBAA177.
[20] X. Chen, C. Clarence Yan, C. Luo, W. Ji, Y. Zhang, and Q. Dai, "Constructing lncRNA functional similarity network based on lncRNA-disease associations and disease semantic similarity," *Scientific Reports 2015 5: 1*, vol. 5, no. 1, pp. 1–12, Jun. 2015, doi:10.1038/srep11338.
[21] N. Peiffer-Smadja *et al.*, "Machine learning for clinical decision support in infectious diseases: a narrative review of current applications," *Clinical Microbiology and Infection*, vol. 26, no. 5, pp. 584–595, May 2020, doi:10.1016/J.CMI.2019.09.009.
[22] S. Gupta *et al.*, "Prioritization of anticancer drugs against a cancer using genomic features of cancer cells: A step towards personalized medicine," *Scientific Reports 2016 6: 1*, vol. 6, no. 1, pp. 1–11, Mar. 2016, doi:10.1038/srep23857.
[23] Q. Wang, Y. H. Feng, J. C. Huang, T. J. Wang, and G. Q. Cheng, "A novel framework for the identification of drug target proteins: Combining stacked auto-encoders with a biased support vector machine," *PLoS One*, vol. 12, no. 4, p. e0176486, Apr. 2017, doi:10.1371/JOURNAL.PONE.0176486.
[24] J. Ha, M. Kambe, and J. Pe, "Data Mining: Concepts and Techniques," *Data Mining: Concepts and Techniques*, pp. 1–703, Jan. 2011, doi:10.1016/C2009-0-61819-5.

[25] G. James, D. Witten, T. Hastie, and R. Tibshirani, "An introduction to statistical learning with applications in R," *Stat Theory Relat Fields*, vol. 6, no. 1, pp. 87–87, Jan. 2022, doi:10.1080/24754269.2021.1980261.

[26] T. Hastie, R. Tibshirani, and J. Friedman, "Springer Series in Statistics The Elements of Statistical Learning Data Mining, Inference, and Prediction".

[27] A. Davies *et al.*, "Advancing mathematics by guiding human intuition with AI," *Nature*, vol. 600, no. 7887, pp. 70–74, Dec. 2021, doi:10.1038/S41586-021-04086-X.

[28] G. E. Hinton and Sejnowski, "Unsupervised learning: Foundations of neural computation," 1999.

[29] T. Ali, S. Ahmed, and M. Aslam, "Artificial Intelligence for Antimicrobial Resistance Prediction: Challenges and Opportunities towards Practical Implementation," *Antibiotics*, vol. 12, no. 3, Mar. 2023, doi:10.3390/ANTIBIOTICS12030523.

[30] I. H. Sarker, A. S. M. Kayes, S. Badsha, H. Alqahtani, P. Watters, and A. Ng, "Cybersecurity data science: an overview from machine learning perspective," *J Big Data*, vol. 7, no. 1, pp. 1–29, Dec. 2020, doi:10.1186/S40537-020-00318-5/FIGURES/3.

[31] J. MacQueen, "Some methods for classification and analysis of multivariate observations," *https: //doi.org/*, vol. 5.1, pp. 281–298, Jan. 1967, Accessed: Apr. 30, 2024. [Online]. Available: https: //projecteuclid.org/ebooks/berkeley-symposium-on-mathematical-statistics-and-probability/Proceedings-of-the-Fifth-Berkeley-Symposium-on-Mathematical-Statistics-and/chapter/Some-methods-for-classification-and-analysis-of-multivariate-observations/bsmsp/1200512992

[32] F. Murtagh and P. Contreras, "Algorithms for hierarchical clustering: an overview, II," *Wiley Interdiscip Rev Data Min Knowl Discov*, vol. 7, no. 6, p. e1219, Nov. 2017, doi:10.1002/WIDM.1219.

[33] Y. Freund and R. E. Schapire, "Experiments with a New Boosting Algorithm," 1996, Accessed: Apr. 30, 2024. [Online]. Available: http: //www.research.att.com/

[34] F. T. Liu, K. M. Ting, and Z. H. Zhou, "Isolation forest," *Proceedings - IEEE International Conference on Data Mining, ICDM*, pp. 413–422, 2008, doi:10.1109/ICDM.2008.17.

[35] S. Wold, K. Esbensen, and P. Geladi, "Principal component analysis," *Chemometrics and Intelligent Laboratory Systems*, vol. 2, no. 1–3, pp. 37–52, Aug. 1987, doi:10.1016/0169-7439(87)80084-9.

[36] T. Hastie, R. Tibshirani, and J. Friedman, "Springer Series in Statistics The Elements of Statistical Learning Data Mining, Inference, and Prediction".

[37] M. L. Puterman, "Markov Decision Processes : Discrete Stochastic Dynamic Programming.," p. 680, 2005, Accessed: Apr. 30, 2024. [Online]. Available: https: //www.wiley.com/en-us/Markov+Decision+Processes%3A+Discrete+Stochastic+Dynamic+Programming-p-9781118625873

[38] R. S. Sutton and A. G. Barto, "Reinforcement Learning: An Introduction Second edition, in progress".

[39] D. Silver *et al.*, "Mastering the game of Go with deep neural networks and tree search," *Nature 2016 529: 7587*, vol. 529, no. 7587, pp. 484–489, Jan. 2016, doi:10.1038/nature16961.

[40] L. P. Kaelbling, M. L. Littman, and A. W. Moore, "Reinforcement Learning: A Survey," *Journal of Artificial Intelligence Research*, vol. 4, pp. 237–285, May 1996, doi:10.1613/JAIR.301.

[41] A. Sakagianni *et al.*, "Using Machine Learning to Predict Antimicrobial Resistance—A Literature Review," *Antibiotics 2023, Vol. 12, Page 452*, vol. 12, no. 3, p. 452, Feb. 2023, doi:10.3390/ANTIBIOTICS12030452.

[42] X. Hao, G. Zhang, and S. Ma, "Deep Learning," *https: //doi.org/10.1142/S1793351X16500045*, vol. 10, no. 3, pp. 417–439, Nov. 2016, doi:10.1142/S1793351X16500045.

[43] "Hassanien A.E (Ed.) - Advanced Machine Learning Technologies and Applications. AMLTA 2020-Springer (2021) | PDF | Artificial Neural Network | Computer Vision." Accessed: Apr. 30, 2024. [Online]. Available: https: //www.scribd.com/document/498066005/Hassanien-A-E-ed-Advanced-machine-learning-technologies-and-applications-AMLTA-2020-Springer-2021

[44] A. Esteva *et al.*, "A guide to deep learning in healthcare," *Nat Med*, vol. 25, no. 1, pp. 24–29, Jan. 2019, doi:10.1038/S41591-018-0316-Z.

[45] C. Chakraborty *et al.*, "Deep learning research should be encouraged more and more in different domains of surgery: An open call –Correspondence," *International Journal of Surgery*, vol. 104, p. 106749, Aug. 2022, doi:10.1016/J.IJSU.2022.106749.

[46] H. Gunasekaran, K. Ramalakshmi, A. Rex Macedo Arokiaraj, S. D. Kanmani, C. Venkatesan, and C. S. G. Dhas, "Analysis of DNA Sequence Classification Using CNN and Hybrid Models," *Comput Math Methods Med*, vol. 2021, 2021, doi:10.1155/2021/1835056.

[47] J. X. Zhang *et al.*, "A deep learning model for predicting next-generation sequencing depth from DNA sequence," *Nature Communications 2021 12: 1*, vol. 12, no. 1, pp. 1–10, Jul. 2021, doi:10.1038/s41467-021-24497-8.

[48] C. Chakraborty *et al.*, "Deep learning research should be encouraged for diagnosis and treatment of antibiotic resistance of microbial infections in treatment associated emergencies in hospitals," *International Journal of Surgery*, vol. 105, p. 106857, Sep. 2022, doi:10.1016/J.IJSU.2022.106857.

6

Artificial Intelligence-Enabled Antimicrobial Stewardship Programs

Gunjan,[1,2] *Ramendra Pati Pandey*[4] and *Chung-Ming Chang*[2,3,*]

Antimicrobial resistance (AMR) presents a formidable global health challenge, necessitating innovative strategies to optimize antibiotic usage. This chapter delves into the integration of Artificial Intelligence (AI) within Antimicrobial Stewardship Programs (ASPs) to enhance clinical decision-making and combat AMR effectively. It outlines the foundational concepts of antimicrobial stewardship (AS), emphasizing its critical role in mitigating AMR and improving patient outcomes. The chapter explores the application of AI technologies, including machine learning, natural language processing, and clinical decision support systems, in analyzing electronic health records and predicting resistance

[1] Graduate Institute of Biomedical Sciences, Chang Gung University, No. 259, Wenhua 1st Road, Guishan Dist. Taoyuan city 33302, Taiwan (R.O.C).
[2] Master & Ph.D. Program in Biotechnology Industry, Chang Gung University, No. 259, Wenhua 1 st Road, Guishan Dist. Taoyuan city 33302, Taiwan (R.O.C).
[3] Department of Medical Biotechnology and Laboratory Science, Chang Gung University, No. 259, Wenhua 1st Road, Guishan Dist. Taoyuan city 33302, Taiwan (R.O.C).
[4] School of Health Sciences and Technology (SoHST), UPES, Bidholi, Dehradun 248007, Uttarakhand, India.
* Corresponding author: cmchang@mail.cgu.edu.tw

patterns. These AI-driven approaches facilitate personalized antibiotic prescribing, real-time surveillance, and proactive intervention strategies. Additionally, the chapter addresses the implementation challenges and future directions of AI-enabled ASPs, underscoring the importance of interdisciplinary collaboration and continuous innovation. By harnessing the potential of AI, healthcare providers can significantly advance antimicrobial stewardship, ensuring the sustainable use of antibiotics and safeguarding public health for future generations.

6.1 Introduction

Antibiotic resistance has emerged rapidly and continues to spread, raising significant concerns about the public health threat of a post-antibiotic world [1]. There's no doubt that the improper or inaccurate use of antimicrobial substances encourages the development and dissemination of A ntimicrobial R esistance (AMR). This not only impacts clinical outcomes but also has significant health and economic ramifications. In contrast to infections in hospitals caused by bacteria susceptible to treatment, this rising prevalence of AMR has become a major concern, as it will not only lead to drug resistance, but it has also increased the risk of super-resistant microorganisms emerging [2]. To address the growth in AMR rates, it is necessary to create an institutional A ntibiotic S tewardship P rogram (ASP) that monitors appropriate antibiotic usage, limits antibiotics, and generates antibiograms.

Utilizing Antimicrobial Stewardship (AMS) interventions is a crucial strategy in addressing the challenge of AMR. ASP implements various measures aimed at optimizing antibiotic usage and preventing inappropriate prescriptions. These include recommending the use of narrower-spectrum antibiotics and shorter durations of antibiotic therapy when deemed appropriate [3]. AMS emphasizes developing and implementing facility-specific clinical practice guidelines. Improving adherence to recommendations can enhance quality of care, but it requires significant human resources. Furthermore, the traditional "passive" way of disseminating clinical practice guidelines is unlikely to effectively modify behavior [4]. AMS interventions are often multifaceted, covering all aspects aimed at optimizing antimicrobial utilization. These include education, diagnostics, guidelines, and prescribing interventions implemented throughout a patient's interaction with healthcare services. It is crucial to survey outcome measures to monitor for positive or negative changes and to guide continuous improvement efforts.

To address the global risk of AMR, ASPs have been implemented to optimize antibiotic use and improve infection management. In recent years, artificial intelligence (AI) has exhibited significant prowess in the realm of AMR control [5]. In both hospital and community settings, ASPs have been shown to be successful in combating AMR [6]. Since hospitals are increasingly moving towards E lectronic H ealth R ecords (EHRs), it is possible to embed various AMS interventions into electronic process flows. In particular, these interventions can focus on the guidelines that govern clinical activities, which are commonly referred to as Clinical decision Support (CDS), inside Electronic Prescribing (eP) software applications. These applications help healthcare workers to follow certain prescribed pathways in their selection and monitoring of antibiotics for patients admitted into any medical facility for treatment, surgery or other purposes [7].

By integrating Artificial intelligence (AI) algorithms into clinical decision-making processes, healthcare providers can easily analyze vast amounts of patient data, identify patterns of resistance, and personalize treatment plans accordingly, leading to more effective outcomes and reduced risks of resistance development [8]. Additionally, AI-enabled predictive analytics can forecast outbreaks and track epidemiological trends, enabling proactive interventions and resource allocation. This symbiotic relationship between AI and antimicrobial stewardship not only enhances patient care but also contributes significantly to global efforts in combating AMR, safeguarding public health for generations to come. This chapter explores various AI applications within antimicrobial stewardship programs (ASP), including machine learning for treatment optimization, natural language processing for data extraction, and predictive analytics for resistance trend forecasting. It highlights AI's potential to revolutionize ASP practices and combat AMR effectively.

6.2 Understanding Antimicrobial Stewardship

6.2.1 Antimicrobial stewardship

Antimicrobial Stewardship (AMS) was initially conceptualized by McGowan, Finland, and others during the 1970s [9]. However, it wasn't until 2007 that the term "AMS" was officially coined by the Infectious Diseases Society of America (IDSA). Subsequently, the Infectious Diseases Society of America (IDSA) and the Society for Healthcare Epidemiology of America (SHEA) released guidelines for establishing institutional programs to bolster AMS [10]. This publication serves as a model for various European guidelines, including the German S3 guideline titled "Strategies for Ensuring Rational Antibiotic Use in Hospitals [11]. Initially AS characterized as structured interventions aimed at enhancing antimicrobial utilization to ensure the selection of appropriate agents, accurate dosage, appropriate route of administration, and optimal therapy duration while safeguarding patient outcomes [12]. Over time, ASPs have evolved beyond mere cost-containment measures to become crucial in optimizing infection treatment in hospitals and reducing antimicrobial selective pressure, thus mitigating the spread of resistance.

While there isn't a single universally accepted definition, most descriptions emphasize that AS involves coordinated initiatives aimed at promoting the proper use of antibiotics. This is done to enhance patient outcomes, mitigate microbial resistance, curb the proliferation of multidrug-resistant organisms, and reduce unnecessary expenses [6]. Additionally, ASP can be defined as holistic approaches designed to maximize the efficacy of antimicrobial agents in healthcare settings. Their significance lies in managing infections caused by multidrug-resistant pathogens. ASPs play a pivotal role in conserving the potency of current antibiotics, mitigating treatment setbacks, forestalling the development of resistance, enhancing patient prognosis, and streamlining healthcare resource allocation. Moreover, ASPs actively contribute to the worldwide endeavor to combat AMR by advocating for judicious antibiotic utilization and curbing the dissemination of multidrug-resistant microorganisms [13].

6.2.2 *Importance of AS in combating AMR*

AMR may arise due to inappropriate utilization of antibiotics, encompassing scenarios such as excessive usage, underuse, abuse, or misuse of antibiotics [14]. Antimicrobial overuse in hospitals, including failure to de-escalate and over prescription of broad-spectrum antibiotics, remains at 50%. ASPs aim to address inappropriate antibiotic usage and AMR [15]. This rapid spread of AMR is worrying, as it could lead to a time when antibiotics no longer work. Prescribing antibiotics plays a big part in causing this resistance, making it crucial for doctors to be careful about when and how they prescribe them [16,17]. Researchers have endeavored to investigate the determinants influencing the prescription of antibiotics by doctors and patients. Their findings indicate that physicians' concerns regarding potential future complications or patient retention, along with factors such as the severity of the patient's condition or allergy history, significantly impact the appropriate prescription of antibiotics [18]. In studies with doctors working in hospitals, they identified some barriers such as: Insufficient assurance in antibiotic guidelines, adherence to established practices, and limited autonomy in decision-making [19]. Therefore, doctors often give too many antibiotics because they're tired of making decisions, want to avoid uncertainty, and feel pressured to meet what patients expect.

6.2.3 Goal of ASP

ASPs are deliberately coordinated initiatives aimed at fostering the judicious and appropriate utilization of antimicrobial agents. ASP can be deployed across various settings where antimicrobials are prescribed, dispensed, or administered, including hospitals, nursing homes, intensive care units, pharmacies, and primary healthcare facilities. The objectives of ASPs are to:

1. Enhance the quality of patient care to promote favorable patient outcomes.
2. Optimize antimicrobial prescribing practices for improved efficacy.
3. Minimize adverse drug reactions associated with antimicrobial agents.
4. Decrease treatment failure rates and mortality among patients.
5. Lower the transmission rate of healthcare-associated infections (HAIs).
6. Mitigate the occurrence of AMR and prevent its spread.
7. Ensure cost-effective therapy and reduce unnecessary healthcare expenditures [20]. (Refer to Fig. 1).

6.2.4 Implementation of AMS across the globe

The implementation of AMS programs has gained momentum globally, driven by the urgent need to address AMR and preserve antibiotic effectiveness. This international movement reflects a collective effort to combat AMR and ensure the sustainability of antibiotic resources [21]. With growing recognition of the importance of AMS, there is a rising demand for detailed guidelines on establishing, executing, and assessing effective AMS initiatives, particularly in developing nations [22]. Education emerges as an important component in translating AMS principles into practical healthcare practice, facilitating widespread

Figure 1: Goal of antimicrobial stewardship program (ASP).

adoption and adherence. Across diverse healthcare systems worldwide, AMS has become a cornerstone strategy in the battle against AMR, underlining its significance in safeguarding public health on a global scale.

6.3 Artificial Intelligence and Antibiotic Stewardship

Current tools and standards for supporting antibiotic prescribing offer valuable guidance but are often too generalized to provide personalized recommendations that effectively balance the immediate risks of undertreatment against the potential risks of overtreatment. This process involves making critical decisions, such as empiric antibiotic prescribing, based on our varying ability to predict the outcome, particularly regarding diagnostic culture susceptibilities. This scenario, characterized by the need for actionable decisions based on variable human judgment, is ideally suited for innovative Machine Learning (ML) methods.

While AI is frequently portrayed as a concept from science fiction, modern algorithms are already integrated into our daily lives. Have you ever noticed how internet advertisements seem tailored specifically to your interests? This is the power of predictive analytics, which employs ML models on extensive datasets to generate personalized predictions and suggestions. Imagine harnessing this same technology to select the most appropriate antibiotic tailored to each individual patient. By leveraging EMR and similar data sources, we can develop predictive models that enhance the accuracy and consistency of empiric antibiotic prescribing, moving beyond traditional educated guess e lectronic patient charts contain a wealth of information that can be harnessed by models, encompassing details such as past infection history, antibiotic susceptibility data, patient-specific symptoms at presentation, medical history, laboratory findings, and imaging results. Currently, researchers utilize this data in large-scale scientific studies to identify independent variables that predict the necessity for specific types of antibiotics in general scenarios. By leveraging appropriate tools, AI models can efficiently analyze similar datasets and, through statistical training, learn to predict responses to specific patient-oriented queries.

For instance, AI models can be trained to determine whether a particular patient has an infection and, if so, whether it's more prudent to adopt a broad-spectrum or targeted treatment approach. With the right input variables and training regimen, these models can quantitatively estimate the probabilities of encountering AMR bacteria in individual patients, a task that humans can only address qualitatively. Furthermore, these predictions can be delivered in real-time, offering the provider's timely guidance during critical decision-making moments. Just like proficient clinicians, AI models should not remain static. Instead, continuously evolving algorithms should adapt to incoming streams of new patient and epidemiological data. With ongoing training, predictions can improve in accuracy and relevance, resulting in a more dynamic guidance tool compared to static guidelines or annual antibiograms work' towards more personalized and effective treatment strategies. In the following sections, we will explore how AI empowers AMS through various tools.

6.4 Electronic Health Records Processing for AI Models to Implement AMS

Electronic health records (EHRs) play a pivotal role in the advancement of AI models designed for AMS implementation. These records, comprising comprehensive patient data ranging from medical history to treatment regimens, serve as invaluable resources for AI algorithms aiming to optimize antibiotic usage. By processing EHRs, AI models can swiftly analyze vast amounts of patient information, identifying patterns, predicting AMR, and suggesting tailored treatment plans. This integration of EHR data with AI facilitates a proactive approach to AMS, ensuring optimal patient care while combat ing AMR effectively. Studies conducted within usual care settings, often referred to as 'point-of-care' trials are made possible by the integration of EHR systems. This pragmatic approach ensures that interventions can be implemented within the context of routine clinical practice, benefiting from the diverse patient populations typical of such settings [23].

It is crucial for these studies to prioritize the human-machine interface, ensuring that interventions designed to achieve behavior change objectives, such as reduced antibiotic prescribing, are meticulously crafted and implemented based on contemporary behavioral science research methods. This emphasis on design and implementation is essential for the success and effectiveness of interventions within EHR systems [24]. One significant advantage of EHR systems is their ability to facilitate a broad spectrum of AMS interventions. By providing streamlined access to electronic medical records (EMRs), these systems enable efficient chart review, simplifying the determination of appropriate antimicrobial drug usage. This capability opens the door to frequent testing and larger scale surveys, enhancing AMS efforts [25]. In the landscape of healthcare delivery, EHR systems serve as comprehensive repositories for patient information, encompassing EMR data, legal documents, and clinical lab results. This wealth of information, gathered from various encounters across care delivery settings, provides physicians with a holistic view of patient health histories. The integration of data from external healthcare providers further enriches this pool of information, ensuring that patient records remain updated and comprehensive [26] (Refer to Fig. 2).

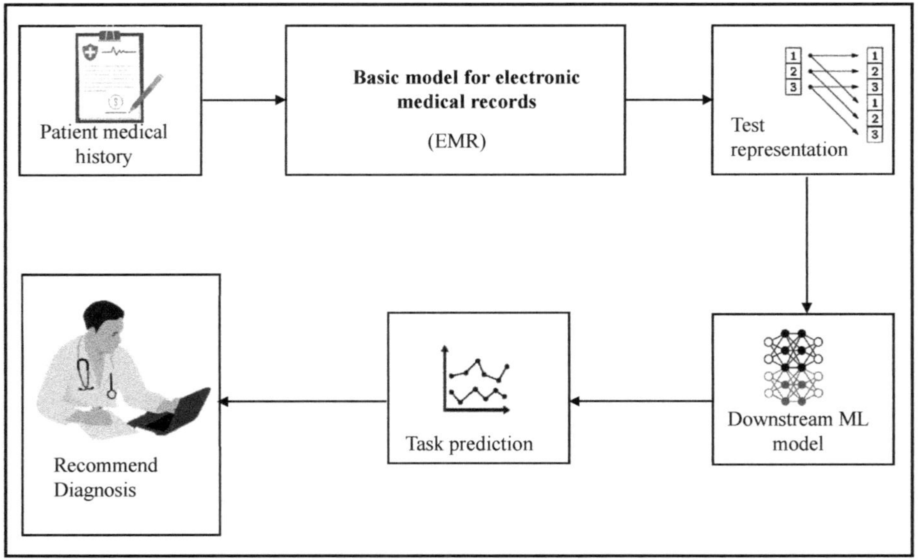

Figure 2: Basic model for EMR designed for AMS implementation.

6.5 Applications of AI in Antimicrobial Stewardship

6.5.1 Machine learning models and neural networks

A similar approach to "algorithmic stewardship" is now acceptable. Algorithms, which are computer-implementable instructions to perform specific tasks, are available for clinical use, including complex AI and ML algorithms and simple rule-based algorithms. More than 50 AI/ML algorithms have been cleared by the US Food and Drug Administration for uses that include identifying intracranial hemorrhage from brain computed tomographic scans and detecting seizures in real time. Algorithms are also used to inform clinical operations, such as predicting which patients will "no show" for scheduled appointments [27].

ML and Neural Networks (NNs) are revolutionizing AMS by providing powerful tools for prediction, surveillance, and decision-making. Within healthcare settings, AMR poses a significant threat, leading to increased morbidity, mortality, and healthcare costs. ML is a division of AI distinguished by its capacity to adapt when confronted with extensive datasets. Unlike expert systems, which are crafted based on human expertise, ML generally operates without direct human intervention (with some exceptions) and endeavors to autonomously discern patterns, akin to the functioning of the human brain. This characteristic renders ML less vulnerable and less reliant on human experts [28]. ML algorithms, also fueled by vast datasets encompassing patient demographics, clinical histories, microbiological data, and antimicrobial usage patterns, can effectively predict the emergence of resistance and identify patients at risk of infection. Neural networks, a subset of ML algorithms inspired by the human brain's structure, excels at recognizing complex patterns within this data, allowing for the early detection of resistant pathogens and the optimization of antimicrobial treatment strategies. By leveraging these technologies, healthcare providers can tailor antimicrobial prescriptions to individual patients, optimizing efficacy while minimizing the risk of resistance development and adverse effects. Furthermore, ML algorithms facilitate real-time surveillance of antimicrobial usage and resistance patterns, enabling healthcare institutions to implement targeted interventions and AMS programs. Numerous studies have showcased the capability of ML algorithms, such as the Gradient Boosted Decision Tree (GBDT), trained on admission data, to forecast the presence of AMR in clinical settings [29]. As these technologies continue to evolve, they hold immense promise for combating antimicrobial resistance, ensuring the sustainable use of antimicrobials, and safeguarding public health on a global scale. However, ethical considerations regarding data privacy, algorithmic bias, and transparency must be carefully addressed to maximize the benefits of ML and neural networks in AMS.

6.5.2 Clinical decision support systems (CDSS)

In recent years, the integration of AI technologies into healthcare has heralded a transformative era in medical practice, fundamentally reshaping our approach to understanding and addressing complex health issues. AI-powered CDSSs driven by sophisticated ML algorithms have emerged as indispensable tools, augmenting the capabilities of healthcare professionals and enhancing clinical decision-making processes. These systems harness the power of big data, drawing upon vast repositories of patient information, including clinical severity metrics, detailed health histories, immunological profiles, prior antibiotic usage patterns, and localized AMR data. By assimilating and analyzing this wealth of data, CDSSs enable clinicians to make more informed and personalized decisions regarding antibiotic selection and prescribing practices, tailored to the unique needs and circumstances of individual patients [30,31].

The potential impact of CDSSs extends far beyond mere assistance in clinical decision-making; these systems hold the promise of revolutionizing medical education and practice on a broad scale. By serving as intelligent guides at the point of care, CDSSs not only enhance the efficiency and accuracy of clinical decisions but also contribute to improved patient outcomes and reduced mortality rates. Empowered by AI-driven insights, healthcare professionals are equipped to navigate the complexities of antibiotic therapy with greater confidence and precision, mitigating the risks of inappropriate treatment choices and AMR [32]. Moreover, CDSSs play a pivotal role in facilitating communication and collaboration among members of the healthcare team, streamlining workflows, and optimizing AMS efforts. Through real-time alerts and comprehensive reports, these systems empower AMS teams to identify patients in need of review based on prescribed therapy, microbiological data, and evolving clinical parameters. This proactive approach not only ensures the timely adjustment of treatment regimens but also promotes judicious antibiotic use, minimizing the risk of overprescribing and the emergence of resistant pathogens [31].

The significance of CDSSs is exemplified by groundbreaking initiatives such as the COMPuterized Antibiotic Stewardship Study

(COMPASS) [33], which represents a paradigm shift in AMS practices. It is a tool of CDSS which was created for the COMPASS trial. This trial is set up as a cluster-randomized, parallel-arm study to assess the impact of a multi-modal computerized AMS intervention [34]. This pioneering study, employing a multi-modal CDSS intervention integrated into the EHR system, seeks to evaluate the impact of advanced decision support tools on antibiotic utilization in diverse clinical settings. By harnessing the collective intelligence of AI and clinical expertise, COMPASS aims to optimize antibiotic prescribing practices, mitigate the burden of AMR, and ultimately improve patient outcomes on a global scale.

Nevertheless, the integration of CDSSs into healthcare ecosystems is not without its challenges and complexities. Concerns regarding the reliability and trustworthiness of AI-derived recommendations, coupled with healthcare professionals' varying degrees of familiarity and acceptance of these technologies, underscore the need for comprehensive research and stakeholder engagement [32,35]. Addressing these challenges requires a nuanced understanding of user perspectives, informed by insights from behavioral science, human factors engineering, and interdisciplinary collaboration. Despite these challenges, the momentum behind CDSS adoption continues to grow, driven by the imperative to harness technology for the betterment of patient care [36]. As healthcare organizations strive to embrace a culture of innovation and evidence-based practice, CDSSs are poised to play an increasingly prominent role in shaping the future of medicine. Whether through regulatory mandates or voluntary adoption, the widespread integration of CDSSs represents a transformative opportunity to enhance patient safety, optimize resource utilization, and advance the delivery of high-quality, data-driven healthcare services (refer to Fig. 3).

6.5.3 Computerized physician order entry (CPOE)

Computerized Physician Order Entry (CPOE) systems play a crucial role in reducing prescription errors by addressing dosing inaccuracies, dosing interval mistakes, and other flaws inherent in traditional methods [36]. These systems empower physicians to assess clinical order information, interpret lab results, prescribe medications, and accurately record their administration—all within a streamlined electronic environment. Moreover, CPOE offers additional functionalities such as checking

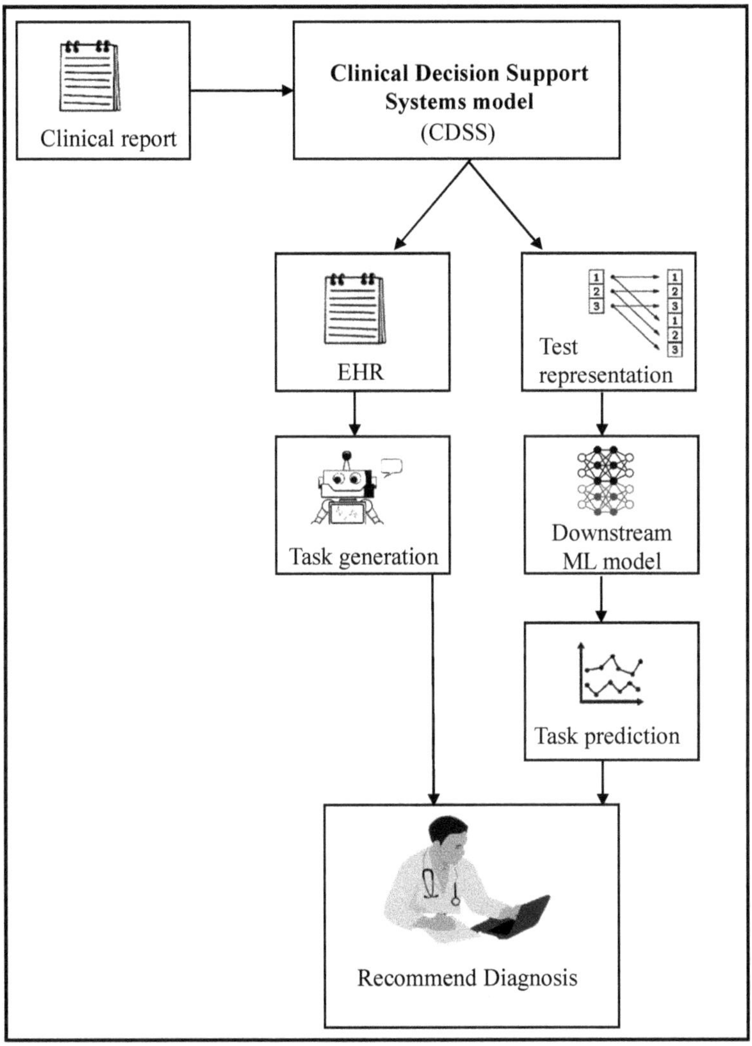

Figure 3: Clinical Decision Support Systems (CDSS) driven by sophisticated ML algorithms.

for drug interactions, providing preventive healthcare reminders, issuing alerts about adverse drug effects, and facilitating proper dosing, especially in patients with renal conditions [37].

Following the implementation of CPOE, the ASP team leverages the system's capabilities to generate reports based on alert notifications. These reports meticulously pinpoint instances where restricted agent alerts are activated, enabling ASP members to thoroughly assess each situation. Utilizing daily alerts, the ASP team verifies prescribers' compliance with approval protocols and provides recommendations on appropriate therapy choices and durations. Previously, such information required manual assessment, significantly extending the time needed for evaluation. In cases where required approval is not sought, ASP team members can promptly follow up with individual prescribers to ensure adherence to ASP protocols [38]. These CPOE-driven interventions within the ASP constitute an integral component, offering the significant advantage of requiring minimal time and staffing resources. Despite the current endorsement of CPOE in the guideline on hospital AMS, there remains a gap in understanding its specific interventions for ASP in pneumonia or Chronic Obstructive Pulmonary Disease (COPD) exacerbation scenarios [39].

CPOE revolutionizes the healthcare system by enabling healthcare providers to electronically prescribe medications and input orders directly into a patient's medical records. One significant advantage of CPOE is the accessibility it offers to healthcare workers, allowing them to efficiently access and review orders and medical charts online, thereby facilitating seamless patient care. However, this innovation comes with its share of challenges, primarily due to the requirement of sophisticated health-care IT systems. Unfortunately, many healthcare facilities lack the infrastructure and resources necessary to implement CPOE effectively, thus limiting its widespread adoption and utilization [40].

6.5.4 Natural language processing (NLP)

Natural Language Processing (NLP) is a multidisciplinary field situated at the intersection of AI and linguistics. Its overarching aim is to automate the analysis of human language [41]. NLP techniques offer a multifaceted approach to ASP, enhancing various aspects of infection control and antibiotic use. By analyzing clinical notes, microbiology reports, and electronic health records, NLP enables the identification of patients at risk of infections or those requiring antimicrobial intervention. This facilitates timely and targeted decision-making by healthcare providers, aligning prescription practices with stewardship guidelines [42]. Moreover, NLP aids in surveillance efforts by automatically monitoring trends in antimicrobial usage and resistance patterns, thereby facilitating proactive interventions to curb the spread of resistant organisms. Additionally, NLP-driven systems can assess adherence to stewardship protocols among healthcare providers, pinpointing instances of inappropriate prescribing and enabling targeted education and intervention strategies. Furthermore, NLP-powered patient engagement tools offer personalized education on responsible antibiotic use, empowering patients to participate in their treatment and reduce unnecessary antibiotic prescriptions. Through these capabilities, NLP plays a pivotal role in optimizing ASP efforts, ultimately improving patient outcomes and combating AMR [43,44] (Fig. 4).

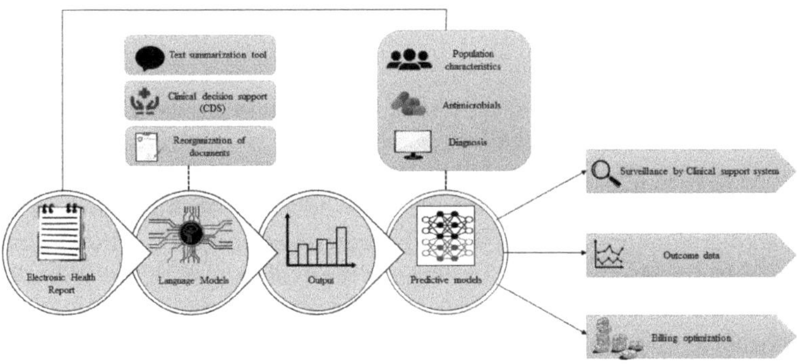

Figure 4: Language models in ASP.

6.6 Challenges and Future Directions

Despite the progress in AMS implementation, several challenges persist, hindering its efficacy and sustainability. One major hurdle is, AI is susceptible to errors and inconsistencies due to data entry issues. Secondly, the complexity of AMR necessitates integrating data from multiple databases, complicating the data cleaning process. Thirdly, EHR data may suffer from nonrandom missing observations, confounding factors, and selection biases [45]. Fourthly, susceptibility data can be affected by changes in clinical breakpoints and shifts in underlying distributions over time, potentially impacting the accuracy of ML models trained on historical data [46]. Lastly, data representations may be institution-specific, hindering generalizability [47]. Addressing these challenges is crucial to ensure that ML CDSS generate accurate recommendations tailored to the local healthcare environment. Careful attention to data quality, integration, and model validation is essential for the successful deployment of ML in combating AMR.

Looking ahead, future directions in AMS must focus on enhancing global collaboration and coordination, promoting the development and dissemination of tailored AMS guidelines, improving access to essential resources, and advancing research in novel diagnostic tools and antimicrobial therapies. Furthermore, efforts should be directed towards integrating AMS into broader health systems strengthening initiatives to ensure sustained impact and mitigate the threat of AMR effectively. By addressing these challenges and embracing innovative approaches, the future of AMS holds promise in safeguarding the efficacy of antimicrobial agents and protecting public health worldwide.

6.7 Conclusion

In conclusion, AMS stands as a vital strategy in the global fight against AMR, yet numerous challenges persist in its implementation and effectiveness. Despite these obstacles, the concerted efforts of healthcare systems worldwide have propelled the adoption of AMS programs, marking significant strides towards preserving antibiotic efficacy and safeguarding public health. To navigate the evolving landscape of AMR, concerted action is imperative. This entails bolstering collaboration, enhancing resources, and prioritizing research and innovation in diagnostics and therapeutics. By addressing these challenges and embracing innovative approaches, the future of AMS holds promise in mitigating the threat of AMR and ensuring the sustainable use of antimicrobial agents for generations to come.

References

[1] C. Huebner, S. Flessa, and N. O. Huebner, "The economic impact of antimicrobial stewardship programmes in hospitals: a systematic literature review," *J. Hosp. Infect.*, vol. 102, no. 4, pp. 369–376, 2019, doi:10.1016/j.jhin.2019.03.002.

[2] M. E. A. de Kraker, P. G. Davey, and H. Grundmann, "Mortality and hospital stay associated with resistant Staphylococcus aureus and Escherichia coli bacteremia: estimating the burden of antibiotic resistance in Europe," *PLoS Med.*, vol. 8, no. 10, Oct. 2011, doi:10.1371/JOURNAL.PMED.1001104.

[3] T. F. Barlam *et al.*, "Implementing an Antibiotic Stewardship Program: Guidelines by the Infectious Diseases Society of America and the Society for Healthcare Epidemiology of America," *Clin. Infect. Dis. An Off. Publ. Infect. Dis. Soc. Am.*, vol. 62, no. 10, p. e51, May 2016, doi:10.1093/CID/CIW118.

[4] S. W. Tu *et al.*, "The SAGE Guideline Model: Achievements and Overview," *J. Am. Med. Inform. Assoc.*, vol. 14, no. 5, p. 589, Sep. 2007, doi:10.1197/JAMIA.M2399.

[5] J. Lv, S. Deng, and L. Zhang, "A review of artificial intelligence applications for antimicrobial resistance," *Biosaf. Heal.*, vol. 3, no. 1, pp. 22–31, Feb. 2021, doi:10.1016/J.BSHEAL.2020.08.003.
[6] P. Davey *et al.*, "Interventions to improve antibiotic prescribing practices for hospital inpatients," *Cochrane database Syst. Rev.*, vol. 2, no. 2, Feb. 2017, doi:10.1002/14651858.CD003543.PUB4.
[7] J. A. Jenkins, S. K. Pontefract, K. Cresswell, R. Williams, A. Sheikh, and J. J. Coleman, "Antimicrobial stewardship using electronic prescribing systems in hospital settings: a scoping review of interventions and outcome measures," *JAC-Antimicrobial Resist.*, vol. 4, no. 3, Jun. 2022, doi:10.1093/JACAMR/DLAC063.
[8] A. Howard *et al.*, "Antimicrobial learning systems: an implementation blueprint for artificial intelligence to tackle antimicrobial resistance," *Lancet Digit. Heal.*, vol. 6, no. 1, pp. e79–e86, Jan. 2024, doi:10.1016/S2589-7500(23)00221-2.
[9] J. E. McGowan and M. Finland, "Usage of antibiotics in a general hospital: effect of requiring justification," *J. Infect. Dis.*, vol. 130, no. 2, pp. 165–168, 1974, doi:10.1093/INFDIS/130.2.165.
[10] T. H. Dellit *et al.*, "Infectious Diseases Society of America and the Society for Healthcare Epidemiology of America guidelines for developing an institutional program to enhance antimicrobial stewardship," *Clin. Infect. Dis.*, vol. 44, no. 2, pp. 159–177, Jan. 2007, doi:10.1086/510393.
[11] K. de With *et al.*, "Strategies to enhance rational use of antibiotics in hospital: a guideline by the German Society for Infectious Diseases," *Infection*, vol. 44, no. 3, pp. 395–439, Jun. 2016, doi:10.1007/S15010-016-0885-Z.
[12] D. Donà *et al.*, "Implementation and impact of pediatric antimicrobial stewardship programs: a systematic scoping review," *Antimicrob. Resist. Infect. Control 2019 91*, vol. 9, no. 1, pp. 1–12, Jan. 2020, doi:10.1186/S13756-019-0659-3.
[13] M. A. A. Majumder *et al.*, "Antimicrobial Stewardship: Fighting Antimicrobial Resistance and Protecting Global Public Health," *Infect. Drug Resist.*, vol. 13, pp. 4713–4738, 2020, doi:10.2147/IDR.S290835.
[14] K. Hand, "Antibiotic stewardship," *Clin. Med.*, vol. 13, no. 5, pp. 499–503, Oct. 2013, doi:10.7861/CLINMEDICINE.13-5-499.
[15] D. Nathwani, D. Varghese, J. Stephens, W. Ansari, S. Martin, and C. Charbonneau, "Value of hospital antimicrobial stewardship programs [ASPs]: A systematic review," *Antimicrob. Resist. Infect. Control*, vol. 8, no. 1, pp. 1–14, 2019, doi:10.1186/s13756-019-0471-0.
[16] World Health Organization, "Global Antimicrobial Resistance and Use Surveillance System (GLASS) Report 2022," *World Heal. Organ.*, no. 8.5.2017, pp. 2003–2005, 2022, Accessed: Apr. 30, 2024. [Online]. Available: www.aging-us.com
[17] I. M. Gould, "Antibiotic resistance: the perfect storm," *Int. J. Antimicrob. Agents*, vol. 34 Suppl 3, no. SUPPL. 3, Aug. 2009, doi:10.1016/S0924-8579(09)70549-7.
[18] P. Lopez-Vazquez, J. M. Vazquez-Lago, and A. Figueiras, "Misprescription of antibiotics in primary care: a critical systematic review of its determinants," *J. Eval. Clin. Pract.*, vol. 18, no. 2, pp. 473–484, Apr. 2012, doi:10.1111/J.1365-2753.2010.01610.X.
[19] J. A. Schouten, M. E. J. L. Hulscher, S. Natsch, B. J. Kullberg, J. W. M. Van Der Meer, and R. P. T. M. Grol, "Barriers to optimal antibiotic use for community-acquired pneumonia at hospitals: a qualitative study," *Qual. Saf. Health Care*, vol. 16, no. 2, pp. 143–149, Apr. 2007, doi:10.1136/QSHC.2005.017327.

[20] "Recommendations for Implementing Antimicrobial Stewardship Programs in Latin America and the Caribbean: Manual for Public Health Decision-Makers," *Recomm. Implement. Antimicrob. Steward. Programs Lat. Am. Caribb. Man. Public Heal. Decis.*, 2018, doi:10.37774/9789275120408.

[21] G. W. Chung, J. E. Wu, C. L. Yeo, D. Chan, and L. Y. Hsu, "Antimicrobial stewardship: a review of prospective audit and feedback systems and an objective evaluation of outcomes," *Virulence*, vol. 4, no. 2, pp. 151–157, Feb. 2013, doi:10.4161/VIRU.21626.

[22] WHO, "Advanced Infection Prevention and Control Training," *World Heal. Organ.*, vol. 2016, no. April, pp. 14–30, 2018.

[23] K. Loudon, S. Treweek, F. Sullivan, P. Donnan, K. E. Thorpe, and M. Zwarenstein, "The PRECIS-2 tool: designing trials that are fit for purpose," *BMJ*, vol. 350, May 2015, doi:10.1136/BMJ.H2147.

[24] L. Yardley, L. Morrison, K. Bradbury, and I. Muller, "The person-based approach to intervention development: Application to digital health-related behavior change interventions," *J. Med. Internet Res.*, vol. 17, no. 1, p. e30, Jan. 2015, doi:10.2196/jmir.4055.

[25] R. Kullar and D. A. Goff, "Transformation of antimicrobial stewardship programs through technology and informatics," *Infect. Dis. Clin. North Am.*, vol. 28, no. 2, pp. 291–300, 2014, doi:10.1016/J.IDC.2014.01.009.

[26] J. Abraham, J. Jaros, I. Ihianle, K. Kochendorfer, and T. Kannampallil, "Impact of EHR-based rounding tools on interactive communication: A prospective observational study," *Int. J. Med. Inform.*, vol. 129, pp. 423–429, Sep. 2019, doi:10.1016/J.IJMEDINF.2019.07.012.

[27] S. Eaneff, Z. Obermeyer, and A. J. Butte, "The Case for Algorithmic Stewardship for Artificial Intelligence and Machine Learning Technologies," *JAMA*, vol. 324, no. 14, pp. 1397–1398, Oct. 2020, doi:10.1001/JAMA.2020.9371.

[28] U. Fanelli *et al.*, "Role of Artificial Intelligence in Fighting Antimicrobial Resistance in Pediatrics," *Antibiot. 2020, Vol. 9, Page 767*, vol. 9, no. 11, p. 767, Nov. 2020, doi:10.3390/ANTIBIOTICS9110767.

[29] M. Cavallaroid, E. Moran, B. Collyer, N. D. Mccarthy, C. Greenid, and M. J. Keeling, "Informing antimicrobial stewardship with explainable AI," 2023, doi:10.1371/journal.pdig.0000162.

[30] B. Rittmann and M. P. Stevens, "Clinical Decision Support Systems and Their Role in Antibiotic Stewardship: a Systematic Review," *Curr. Infect. Dis. Rep.*, vol. 21, no. 8, pp. 1–12, Aug. 2019, doi:10.1007/S11908-019-0683-8/FIGURES/3.

[31] "Antimicrobial Stewardship Strategy: CDSS/CPOE Antimicrobial Stewardship Strategy: Clinical decision support systems/computerized physician order entry", Accessed: Apr. 30, 2024. [Online]. Available: http: //www.safetyandquality.gov.au/wp-content/uploads/2011/01/Antimicrobial-stewardship-in-

[32] T. M. Rawson *et al.*, "A systematic review of clinical decision support systems for antimicrobial management: are we failing to investigate these interventions appropriately?," *Clin. Microbiol. Infect.*, vol. 23, no. 8, pp. 524–532, Aug. 2017, doi:10.1016/J.CMI.2017.02.028.

[33] G. Catho *et al.*, "How to Develop and Implement a Computerized Decision Support System Integrated for Antimicrobial Stewardship? Experiences From Two Swiss Hospital Systems," *Front. Digit. Heal.*, vol. 2, p. 583390, Feb. 2021, doi:10.3389/FDGTH.2020.583390/BIBTEX.

[34] G. Catho *et al.*, “Study protocol for a multicentre, cluster randomised, superiority trial evaluating the impact of computerised decision support, audit and feedback on antibiotic use: the COMPuterized Antibiotic Stewardship Study (COMPASS),” *BMJ Open*, vol. 8, no. 6, Jun. 2018, doi:10.1136/BMJOPEN-2018-022666.

[35] K. Kawamoto, C. A. Houlihan, E. A. Balas, and D. F. Lobach, “Improving clinical practice using clinical decision support systems: a systematic review of trials to identify features critical to success,” *BMJ*, vol. 330, no. 7494, pp. 765–768, Apr. 2005, doi:10.1136/BMJ.38398.500764.8F.

[36] F. Albahar, R. K. Abu-Farha, O. Y. Alshogran, H. Alhamad, C. E. Curtis, and J. F. Marriott, “Healthcare Professionals’ Perceptions, Barriers, and Facilitators towards Adopting Computerised Clinical Decision Support Systems in Antimicrobial Stewardship in Jordanian Hospitals,” *Healthc. 2023, Vol. 11, Page 836*, vol. 11, no. 6, p. 836, Mar. 2023, doi:10.3390/HEALTHCARE11060836.

[37] J. Metcalfe, A. Lam, S. S. H. Lam, J. M. de Clifford, and P. Schramm, “Impact of the introduction of computerised physician order entry (CPOE) on the surveillance of restricted antimicrobials and compliance with policy,” *J. Pharm. Pract. Res.*, vol. 47, no. 3, pp. 200–206, 2017, doi:10.1002/jppr.1227.

[38] E. C. Webber, H. M. Warhurst, S. S. Smith, E. G. Cox, A. S. Crumby, and K. R. Nichols, “Conversion of a single-facility pediatric antimicrobial stewardship program to multi-facility application with computerized provider order entry and clinical decision support,” *Appl. Clin. Inform.*, vol. 4, no. 4, pp. 556–568, 2013, doi:10.4338/ACI-2013-07-RA-0054.

[39] S. K. Gohil *et al.*, “Stewardship Prompts to Improve Antibiotic Selection for Urinary Tract Infection: The INSPIRE Randomized Clinical Trial,” *JAMA*, Apr. 2024, doi:10.1001/JAMA.2024.6259.

[40] A. W. H. O. P. Toolkit, *Antimicrobial stewardship programmes in health-care facilities in low- and middle-income countries: a WHO practical toolkit*, vol. 1, no. 3. 2019. doi:10.1093/jacamr/dlz072.

[41] P. M. Nadkarni, L. Ohno-Machado, and W. W. Chapman, “Natural language processing: an introduction,” *J. Am. Med. Inform. Assoc.*, vol. 18, no. 5, pp. 544–551, Sep. 2011, doi:10.1136/AMIAJNL-2011-000464.

[42] E. A. Özçelik *et al.*, “A comparative assessment of action plans on antimicrobial resistance from OECD and G20 countries using natural language processing,” *Health Policy (New. York).*, vol. 126, no. 6, pp. 522–533, Jun. 2022, doi:10.1016/J.HEALTHPOL.2022.03.011.

[43] H. Atluri, B. Siva, P. Thummisetti, and I. Bgsu, “Enhancing Antibiotic Prescribing in Urgent Care by Leveraging Large Language Models for Optimized Clinical Decision Support,” *www.irjmets.com @International Res. J. Mod. Eng.*, 2009, doi:10.56726/IRJMETS48495.

[44] D. Baggio *et al.*, “Closing the Gap in Surveillance and Audit of Invasive Mold Diseases for Antifungal Stewardship Using Machine Learning,” *J. Clin. Med. 2019, Vol. 8, Page 1390*, vol. 8, no. 9, p. 1390, Sep. 2019, doi:10.3390/JCM8091390.

[45] S. Mani *et al.*, “Medical decision support using machine learning for early detection of late-onset neonatal sepsis,” *J. Am. Med. Informatics Assoc.*, vol. 21, no. 2, pp. 326–336, 2014, doi:10.1136/amiajnl-2013-001854.

[46] A. Subbaswamy and S. Saria, "From development to deployment: dataset shift, causality, and shift-stable models in health AI," *Biostatistics*, vol. 21, no. 2, pp. 345–352, Apr. 2020, doi:10.1093/BIOSTATISTICS/KXZ041.

[47] D. Ghosh *et al.*, "Machine learning based prediction of antibiotic sensitivity in patients with critical illness," *medRxiv*, p. 19007153, Sep. 2019, doi:10.1101/19007153.

7

Big Data Analytics and Antimicrobial Resistance

Saif Ali and *Subhajit Basu**

Antimicrobial resistance is a significant global health challenge, where microorganisms are increasingly resistant to antibiotics due to factors like genetic mutations, horizontal gene transfer, and improper antibiotic usage. This resistance complicates the treatment of common infections. The use of big data analytics has become essential in the battle against the growing threat posed by antimicrobial resistance. In order to address this global health concern, modern analytics can improve surveillance, diagnosis, and research efforts by integrating different data sources, including genomic information, prescribing data, and microbiology results. The field of antimicrobial identification has taken a new turn due to the merging of artificial intelligence (AI), particularly deep learning and machine learning. In this chapter, we mostly look into big data analytics tool utiliz ation. Combining different data sources makes it easier to conduct surveillance and monitoring and allows for the immediate identification of AMR trends, facilitating the early discovery of novel resistance patterns.

7.1 Introduction

One of the major threats to world health is Antimicrobial Resistance (AMR) that requires innovative solutions. AMR significantly risks public

University of Petroleum and Energy Studies, Dehradun, India.
* Corresponding author

health around the world by making many of the current antibiotics useless against common bacterial diseases [1]. Microorganisms can withstand drugs meant to eradicate them, situations arise in which common diseases might become fatal. Agricultural practices, poor sanitation, overuse and misuse of medicines in humans and animals, and other variables all play a part in the complicated issue of the emergence and spread of AMR [2]. Antimicrobial resistance occurs when microorganisms, such as bacteria and fungi evolve mechanisms to withstand the effects of antimicrobial agents, including antibiotics, antivirals, antifungals, and antiparasitics. This resistance can arise through various mechanisms, including genetic mutations, horizontal gene transfer, and selective pressure exerted by antimicrobial use [3].

Microorganisms can become resistant to antimicrobial medications by spontaneous mutations in their genetic makeup, especially in genes encoding targets of antimicrobial activity or in genes involved in drug uptake or efflux. These mutations may alter the structure or function of cellular components, such as enzymes or transport proteins, leading to reduced drug susceptibility [5].

Horizontal gene transfer is a key mechanism by which bacteria acquire resistance genes from other microorganisms in their environment. This process involves the transfer of genetic material, including plasmids, transposons, and integrons between different bacterial strains or species. Resistance genes carried on mobile genetic elements can spread rapidly within microbial populations, contributing to the dissemination of antimicrobial resistance [4]. Bacterial resistance mechanisms encompass a diverse array of strategies, including enzymatic inactivation of drugs (e.g., beta-lactamases), alteration of drug targets (e.g., mutation of bacterial ribosomes or DNA gyrase), and efflux pumps that actively extrude antimicrobial agents from bacterial cells [6]. Additionally, bacteria can develop biofilm formation, which provides protection against antibiotics and host immune responses. Antifungal resistance mechanisms include alterations in the structure or expression of fungal cell wall components (e.g., glucans, chitin), mutations in genes encoding drug targets (e.g., ergosterol biosynthesis enzymes), and upregulation of efflux pumps that expel antifungal agents from fungal cells (Fisher et al., 2018) [29].

The emergence of multidrug-resistant pathogens has necessitated innovative approaches to combat this pressing issue. In recent years, big data analytics has emerged as a powerful tool in the fight against antimicrobial resistance, offering insights into the dynamics of resistance development, transmission pathways, and potential interventions. In order to derive valuable insights and patterns, big data analytics entails gathering, processing, and analyz ing massive amounts of data. Data from clinical, genetic, epidemiological, and environmental sources can all be integrated with it. We can obtain a thorough grasp of the mechanisms underlying antibiotic resistance by merging these databases. In order to forecast the emergence and spread of antibiotic resistance, it also enables researchers to create predictive models. Early diagnosis and intervention are made possible by these models, which use machine learning algorithms to find patterns and trends in data.

Collecting data from healthcare facilities, laboratories, and surveillance networks, researchers can track the prevalence of resistant pathogens and identify emerging hotspots [7]. Additionally, big data analytics can aid in tracking the spread of antimicrobial-resistant pathogens, identifying clusters or outbreaks, and informing strategies for infection control and prevention (Bansal et al., 2016) [30]. Furthermore, data analytics can also support the development of predictive models to identify individuals at high risk for developing antimicrobial resistance and guide targeted interventions. This can ultimately lead to more effective treatment protocols and the preservation of vital antimicrobial drugs. The use of big data analytics in combating antimicrobial resistance has the potential to revolutionize global healthcare and improve patient outcomes by identifying patterns of overuse and inappropriate use of antimicrobials, tracking the spread of resistant pathogens, and guiding targeted interventions. Therefore, this chapter explores the role of big data analytics in addressing antimicrobial resistance, including its applications, challenges, and future prospects.

7.2 Comparative Studies of AMR using Traditional and Tools based on Data Analytics

In 2023, Luz et al. conducted a comparative study between conventional data analysis and reporting techniques and a recently devised way for AMR data processing. Evaluating usability, accuracy, and efficiency in a clinical context was the main goal. The results of blood culture tests were delivered to ten specialists who work with AMR data, and they were instructed to analyze the data using both old and modern software tools. They carried out tasks using both approaches, answered a questionnaire, and assessed usability. Accuracy, time spent, and task completion were all factors in the data analysis.

Time Efficiency: (93.7 to 22.4 minutes; P < 0.001) was the mean amount of time spent preparing AMR reports, a considerable reduction.

Task Completion and Accuracy: There was an improvement in average task completion from 56% to 96% (P < 0.05). The percentage of right responses rose from 37.9% to 97.9% (P<0.001).

Usability: According to the System Usability Scale (SUS), the new tools' median score was 83.8, which is considered to be good.

Utilizing open-source software tools for AMR data analysis and reporting was shown to be efficacious, efficient, and accurate by the study. The study emphasized the drawbacks of conventional reporting instruments and the requirement for better approaches to AMR data reporting and analysis. It also offers insightful information on the difficulties associated with AMR data reporting and analysis in clinical settings, as well as the possible advantages of implementing novel strategies to overcome these difficulties.

7.3 Role of Artificial Intelligence in Antimicrobial Resistance

Artificial intelligence (AI) stands at the forefront of modern computer science, striving to replicate human intelligence and tackle tasks ranging from visual perception to natural language processing [8] Obermeyer and Emanuel, 2016 [31]. Fueling its evolution are the abundance of data gleaned from Electronic Health Records (EHR) and leaps in computational power. Machine L earning (ML) and Neural Networks (NN), complex mathematical functions at AI's core, have seen particular advancement, especially with the rise of Deep Neural Network (DNN) architectures. ML, unlike expert systems, autonomously learns from data, making it less reliant on human input and more adaptable. NNs, inspired by biological neural networks, discern patterns within data through interconnected processing units. DNNs, with multiple layers of processing units, excel at understanding vast datasets, continually improving with more data Waldstein et al., 2018; [11]. In the medical realm, where patient care is increasingly complex and generates an overwhelming volume of data, AI holds promise in revolutionizing clinical decision-making [10]. Despite this potential, many healthcare professionals remain unaware of AI's utility, often relying solely on personal experience and guidelines. Amidst this backdrop, the exploration of AI's role in combat ing antimicrobial resistance (AMR) emerges as a critical endeavor, particularly in pediatric care [9]. The exponential growth of medical information underscores the need for AI's assistance in managing data overload, with studies suggesting that the amount of medical data produced annually far surpasses what an individual could consume in a lifetime.

7.4 Tools used for Antimicrobial Resistance

Methods like sequence assembly [12,13,14] and read-based techniques [15,16] are employed to discover how bacteria develop resistance to antibiotics. These methods are good at finding known resistance genes, but they can also make mistakes and identify genes that aren't actually responsible for resistance. Machine-learning, a type of computer technique, offers another way to find these resistance genes. Instead of just comparing sequences, machine-learning looks for specific characteristics in the genes that are unique to antibiotic resistance. Some methods have been developed using machine-learning to find new resistance genes, but they often use only a few gene features for predictions. Also, they don't always filter out irrelevant features, which can make the predictions less accurate.

Chowdhary et al. developed a software program called "Prediction of Antimicrobial Resistance via Game Theory " (PARGT). It features a user-friendly graphical interface designed to identify antimicrobial-resistance genes in both Gram-positive and Gram-negative bacteria. PARGT integrates various tools and scripts needed to identify protein features and generate feature subsets using the "game theoretic dynamic weighting based feature evaluation" (GTDWFE) algorithm. It allows users to predict resistance genes in different types of bacteria. Users can test individual sequences or entire genomes for specific resistance genes. Additionally, PARGT enables users to update the training set with newly confirmed AMR or non-AMR sequences and reset the training data to its original state.

7.5 Access, Watch, Reserve Tool (AWaRe)

The "AWaRe Tool" promotes the use of narrow-spectrum antibiotics for common infections, reserving broad-spectrum antibiotics for the most challenging cases. It offers recommendations for 21 prevalent infections and classifies antibiotics into three groups, aligning with WHO's.

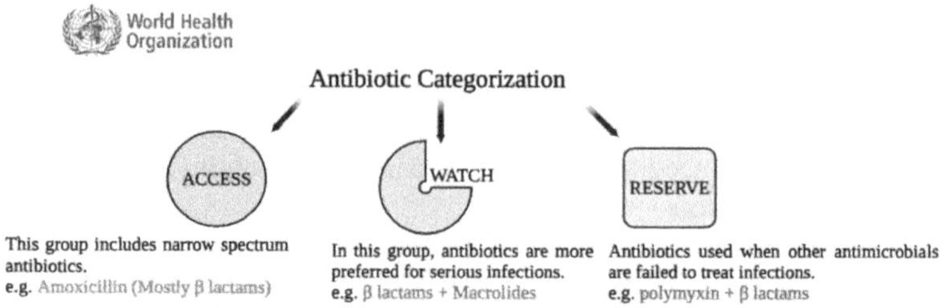

Figure 1: Antibiotic characterization by WHO into three groups. guidance for antibiotic use [18,17].

7.6 Whole Genome Sequencing (WGS)

Whole Genome Sequencing (WGS) is a technique used to analyze the genomes of both prokaryotic and eukaryotic organisms, including individual isolates and simultaneous sequencing of all organisms in a sample (metagenomics). Various sequencing technologies which are able to generate short-read and long-read sequences, are available [19]. In microbiology, WGS is valuable for mapping bacterial isolates' epidemiology in clinical outbreaks and identifying potentially pathogenic strains in food and environmental samples [20]. Sequencing is used to identify antimicrobial resistance (AMR) genes in isolates, guiding antibiotic treatment decisions [21]. Many bioinformatic software and pipelines predict AMR genes in genomic and metagenomic data, involving direct analysis or assembly into bacterial chromosomes, contigs, and plasmids. Benchmarking these algorithms on common datasets enhances confidence in WGS results interpretation.

7.7 MEGARes

Appropriate resources are needed to properly analyze large-scale data as techniques for identifying and characterizing antimicrobial resistance shift from targeted culture and polymerase chain reaction to high throughput metagenomics. Currently available datasets of antimicrobial resistance are made with thorough annotations and intended for smaller-scale functional gene profiling. These characteristics, however, make it more difficult to analyze ecological sequence databases of a huge size, like those produced by metagenomics surveillance. MEGARes is an antimicrobial resistance (AMR) database that has been meticulously chosen for use in sequence categorization, statistical analysis, and high-throughput sequencing (HTS) data processing. MEGARes ensures dependability by rigorously verifying each entry manually for accuracy at the protein and nucleotide levels [22].

7.8 Resfinder

Accurately identifying resistant bacteria and antimicrobial resistance genes (ARGs) is crucial for understanding how antimicrobial resistance (AMR) evolves and emerges, as well as for guiding treatment decisions [24]. ResFinder is a free online tool designed for this purpose, analyzing next-generation sequencing data to predict bacterial phenotypes (Schwarz et al., 2020). It categorizes ARGs by antimicrobial class and provides detailed gene descriptions, including names, accession numbers, associated antimicrobials, and resistance mechanisms.

Meanwhile, the PointFinder database focuses on mutations in specific genes across various bacterial species that lead to antimicrobial resistance. It covers substitutions, deletions, and insertions in coding regions, promoters, or ribosomal RNA. Each mutation is linked to relevant articles, resistance mechanisms, and additional mutations necessary for resistance [23].

Both databases are maintained by researchers who rigorously review published articles to ensure the inclusion of novel resistance genes or mutations.

7.9 CARD

Comprehensive Antibiotic Resistant Database (CARD) is a meticulously curated repository that houses comprehensive reference data concerning the molecular mechanisms underlying antimicrobial resistance (AMR), with particular focus on genes, proteins, and mutations associated with AMR. Employing an ontological framework, CARD is structured around models and encompasses a wide array of AMR drug categories and resistance mechanisms, including intrinsic, mutation-driven, and acquired resistance. Fundamentally reliant on the Antibiotic Resistance Ontology (ARO), CARD employs a custom-built, interconnected, and hierarchical controlled vocabulary to facilitate advanced data organization and sharing. Its architecture is conducive to the development of innovative genomic analysis tools, exemplified by the Resistance Gene Identifier (RGI), which enables prediction of resistomes directly from raw genome sequences [26].

7.10 Applications of Big Data Analytics in Antimicrobial Resistance

Big data analytics offers several applications in effectively addressing the challenge of AMR. Firstly, it makes surveillance and monitoring easier by combining various data sources to identify AMR trends instantly, which enables the early detection of new resistance patterns. Second, advanced data analytics help build rapid diagnostic tests that detect resistant bacteria in a timely manner, directing physicians to provide the right medicines at the onset to fight resistance. Thirdly, the creation of novel antimicrobial treatments and methods to reduce resistance is influenced by big data on microbial genetics, resistance mechanisms, and patient outcomes.

Comprehensive data on antibiotic use and resistance-promoting factors can lend support to evidence-based guidelines and policies that encourage prudent antimicrobial prescribing. Additionally, big data analytics can forecast how each patient will react to antibiotics, allowing for individualized treatment plans that reduce the chance of resistance emerging. Big data analytics improves global monitoring and response activities by promoting global collaboration and data sharing, which gives a complete picture of AMR trends globally. Innovative solutions arise by expanding the scope of AMR surveillance and study through the integration of novel data sources such as social media, environmental monitoring, and genomic data. Furthermore, ethical antibiotic use and behavio ral modifications to reduce resistance are encouraged by using big data analytics to improve public awareness and education about AMR. These examples show the variety of ways that big data analytics might be applied to tackle the challenging problem of antibiotic resistance.

7.11 Importance of Big Data Analytics to Combat AMR

Big data analytics integrates various data sources like microbiology results, prescribing data, and mortality statistics, providing a comprehensive and real-time view of AMR trends for more effective surveillance and early detection of emerging resistance. Advanced analytics aid in developing rapid diagnostic tests for identifying resistant pathogens swiftly, enabling clinicians to prescribe appropriate antibiotics from the outset, thus preventing the use of ineffective drugs that may exacerbate resistance [28]. Leveraging big data on microbial genomics, resistance mechanisms, and patient outcomes uncovers insights guiding the development of novel antimicrobial therapies and strategies for mitigating resistance. Comprehensive data on antibiotic use and its impact on resistance supports evidence-based guidelines and policies to promote judicious antimicrobial prescribing. Big data analytics predicts individual patient responses to antibiotics, facilitating tailored treatment approaches that minimize resistance risk.

7.12 Challenges of using Big Data Analytics in AMR

Combining different types of data on antimicrobial resistance, like microbiology records, prescriptions, and mortality figures, is tough because they're usually kept separate. Keeping patient data safe while using it for AMR research is a big challenge. Experts from various fields need to work together to fight AMR using big data, but it's hard to get them to collaborate effectively. Predicting how AMR will spread is hard because the data isn't perfect and there are many factors involved. High staff turnover makes it even harder for big healthcare organizations to do tasks like preparing samples for genome sequencing, so they need to automate these tasks [27]. When collecting data, it's important to focus on specific goals rather than trying to gather as much data as possible. Although big data analytics has the potential to improve AMR research and control, there are many technical, organizational, and regulatory challenges that need to be dealt with first.

7.13 Future Perspectives

The future prospects of big data analytics in combating antimicrobial resistance (AMR) hold great promise, offering significant potential for bolstering surveillance, research, and policy development. This involves establishing seamless integration and interoperability among diverse data sources, such as microbiology findings, prescription records, and mortality statistics, to gain a comprehensive understanding of AMR trends and patterns. Machine learning and artificial intelligence can be harnessed to analyze extensive datasets, identify trends, and forecast the spread of AMR, thereby informing more efficient surveillance, diagnostics, and treatment strategies. Big data analytics can also be employed to craft personalized treatment approaches and precise public health strategies, considering individual patient responses to antibiotics and environmental factors impacting AMR. Moreover, fostering global collaboration and data sharing is crucial to creating a comprehensive, real-time perspective of AMR trends worldwide, facilitating more effective international surveillance and response efforts.

Addressing data challenges such as quality, standardization, and security is essential to ensure the credibility and dependability of AMR data analytics. Additionally, integrating innovative data sources like genomic data, environmental monitoring, and social media can broaden the scope of AMR surveillance and research. Developing new analytics tools and methodologies capable of handling the complexity and scale of AMR data, such as machine learning algorithms and data visualization techniques, is imperative. Furthermore, utilizing big data analytics can enhance public awareness and education about AMR, advocating responsible antibiotic usage, and promoting behavioral changes to mitigate AMR spread.

References

[1] D. Chinemerem Nwobodo, M.C. Ugwu, C. Oliseloke Anie, M.T. Al-Ouqaili, J. Chinedu Ikem, U. Victor Chigozie, and M. Saki. Antibiotic resistance: the challenges and some emerging strategies for tackling a global menace J. Clin. Lab. Anal., 36 (9) (2022), p. e24655.

[2] E. Palma, B. Tilocca, and P. Roncada.. Antimicrobial Resistance in Veterinary Medicine: An Overview. International journal of molecular sciences, 21(6), 1914, 2020. https: //doi.org/10.3390/ijms21061914

[3] A. H. Holmes, L. S. P. Moore, A. Sundsfjord, M. Steinbakk, S. Regmi, A. Karkey, and P.J. Guerin. Understanding the mechanisms and drivers of antimicrobial resistance. The Lancet, 387(10014), 176–187, 2016. https: //doi.org/10.1016/S0140-6736(15)00473-0

[4] J. M. Blair, M. A. Webber, A. J. Baylay, D. O. Ogbolu, and and L. J. Piddock. Molecular mechanisms of antibiotic resistance. Nature Reviews Microbiology, 13(1), 42–51 2015.

[5] J. M. Blair, and L. J. Piddock. How to measure export via bacterial multidrug resistance efflux pumps. MBio, 7(4), e00840–16, 2016.

[6] H. Nikaido. Multidrug resistance in bacteria. Annual Review of Biochemistry, 78, 119–146, 2009.

[7] E. Vayena, J. Dzenowagis, J. S. Brownstein, and A. Sheikh. Policy implications of big data in the health sector. Bulletin of the World Health Organization, 96(1), 66–68, 2018. https: //doi.org/10.2471/BLT.17.197426

[8] S. Russell, and P. Norvig. Artificial Intelligence: A Modern Approach. 2009. Pearson.

[9] U. Fanelli, M. Pappalardo, V. Chinè, P. Gismondi, C. Neglia, A. Argentiero, A. Calderaro, A., A. Prati, and S. Esposito. Role of Artificial Intelligence in Fighting Antimicrobial Resistance in Pediatrics. Antibiotics (Basel, Switzerland), 9(11), 767, 2020. https: //doi.org/10.3390/antibiotics9110767

[10] Y. LeCun, Y. Bengio, G. Hinton. Deep learning. Nature, 521(7553), 436–444, 2015.

[11] U. Schmidt-Erfurth, H. Bogunović, A. Sadeghipour, T. Schlegl, G. Langs, B. S. Gerendas, A. Osborne, and S. M. Waldstein. Machine learning to analyze the prognostic value of current imaging biomarkers in neovascular age-related macular degeneration. Ophthalmol. Retin. 2018; 2: 24–30. doi: 10.1016/j.oret.2017.03.015.

[12] U. Fanelli, V. Chiné., M. Pappalardo, P. Gismondi, and S. Esposito. Improving the quality of hospital antibiotic use: Impact on multidrug-resistant bacterial infections in children. Front. Pharmacol. 2020; 11: 745. doi: 10.3389/fphar.2020.00745.

[13] E. Zankari, *et al.* Identification of acquired antimicrobial resistance genes. *J. Antimicrob. Chemother.* **67**, 2640–2644, 2012.

[14] B. Jia. *et* al. Card 2017: Expansion and model-centric curation of the comprehensive antibiotic resistance database. *Nucleic Acids Res.* gkw1004, 2016.

[15] M. Inouye. *et al.* Srst2: Rapid genomic surveillance for public health and hospital microbiology labs. *Genome Med.* **6**, 90, 2014.

[16] W. Rowe. *et al.* Search engine for antimicrobial resistance: A cloud compatible pipeline and web interface for rapidly detecting antimicrobial resistance genes directly from sequence data. *PLoS One* **10**, e0133492, 2015.

[17] World Health Organization. WHO Antibiotic Categorization. [Last accessed on 2024 April 25].

[18] P. K. Goyal, A. Semwal, A. Prakash, and B. Medhi. Emerging antimicrobial resistance and newer tools to address the resistance. Indian journal of pharmacology, 51(5), 291–295, 2019. https: //doi.org/10.4103/ijp.IJP_607_19

[19] S.M. Karst. *et al.* High-accuracy long-read amplicon sequences using unique molecular identifiers with Nanopore or PacBio sequencing. *Nat. Methods* **18**, 165–169, 2021.

[20] S. R. Simar, B. M. Hanson, and C. A. Arias. Techniques in bacterial strain typing: past, present, and future. *Curr. Opin. Infect. Dis.* **34**, 339–345, 2021.

[21] A. L. Cooper. *et al.* Systematic evaluation of whole genome sequence-based predictions of *Salmonella* serotype and antimicrobial resistance. *Front. Microbiol.* **11**, 549, 2020.

[22] S. M. Lakin, C. Dean, N. R. Noyes, A. Dettenwanger, A. S. Ross, E. Doster, P. Rovira, Z. Abdo, K. L. Jones, J. Ruiz, K. E. Belk, P. S. Morley, and C. Boucher. MEGARes: an antimicrobial resistance database for high throughput sequencing. Nucleic acids research, 45(D1), D574–D580, 2017. https: //doi.org/10.1093/nar/gkw1009

[23] A. F. Florensa, R. S. Kaas, P. T. L. C. Clausen, D. Aytan-Aktug, and F. M. Aarestrup. ResFinder—an open online resource for identification of antimicrobial resistance genes in next-generation sequencing data and prediction of phenotypes from genotypes. Microbial genomics, 8(1), 000748, 2022. https: //doi.org/10.1099/mgen.0.000748

[24] World Health Organization Antimicrobial resistance: global report on surveillance. 2014. http: //apps.who.int/iris/bitstream/10665/112642/1/9789241564748_eng.pdf

[25] V. Bortolaia, R. S. Kaas, E. Ruppe, M. C. Roberts, S. Schwarz. *et al.* ResFinder 4.0 for predictions of phenotypes from genotypes. J Antimicrob Chemother. 2020; 75: 3491–3500. doi: 10.1093/jac/dkaa345.

[26] B. Jia, A. Raphenya, B. Alcock, N. Waglechner, P. Guo, K. Tsang, B. Lago, B. Dave, S. Pereira, A. Sharma, S. Doshi, M. Courtot, R. Lo, L. Williams, J. Frye, T. Elsayegh, D. Sardar, E. Westman, A. Pawlowski, T. Johnson, F. Brinkman, G. Wright, and A. McArthur. CARD 2017: expansion and model-centric curation of

the comprehensive antibiotic resistance database. Nucleic Acids Research, 45, D566–D573, 2016. https: //doi.org/10.1093/nar/gkw1004.

[27] S. Pei, S. Blumberg, J. C. Vega, T. Robin, Y. Zhang, R. J. Medford, B. Adhikari, J. Shaman, and CDC MIND-Healthcare Program. Challenges in Forecasting Antimicrobial Resistance. Emerging infectious diseases, 29(4), 679–685, 2023. https: //doi.org/10.3201/eid2904.221552

[28] T. F. O'Brien, A. Clark, R. Peters, and J. Stelling. Why surveillance of antimicrobial resistance needs to be automated and comprehensive. Journal of global antimicrobial resistance, 17, 8–15, 2019. https: //doi.org/10.1016/j.jgar.2018.10.011

[29] M. C. Fisher, N. J. Hawkins, D. Sanglard, and S. J. Gurr, "Worldwide emergence of resistance to antifungal drugs challenges human health and food security," Science, vol. 360, no. 6390, pp. 739–742, May 2018. [Online]. Available: https://doi.org/10.1126/science.aap7999

[30] S. Bansal, G. Chowell, L. Simonsen, A. Vespignani, and C. Viboud, "Big data for infectious disease surveillance and modeling," The Journal of Infectious Diseases, vol. 214, no. suppl_4, pp. S375–S379, Dec. 2016. [Online]. Available: https://doi.org/10.1093/infdis/jiw400

[31] Z. Obermeyer and E. J. Emanuel, "Predicting the future - big data, machine learning, and clinical medicine," The New England Journal of Medicine, vol. 375, no. 13, pp. 1216–1219, Sep. 2016. [Online]. Available: https://doi.org/10.1056/NEJMp1606181

[32] A. S. Chowdhury, D. R. Call, and S. L. Broschat, "PARGT: a software tool for predicting antimicrobial resistance in bacteria," Scientific Reports, vol. 10, p. 11033, Jul. 2020. [Online]. Available: https://doi.org/10.1038/s41598-020-67949-9

8

Artificial Intelligence in Veterinary Medicine and Antimicrobial Resistance

Siya Sharma and *Subhajit Basu*

A serious threat to animal welfare, health, and food security is antimicrobial resistance (AMR). To mitigate the emergence and spread of resistance infections, veterinarians play a crucial role in using antimicrobials appropriately. Artificial intelligence (AI) has become a viable technology for fighting antimicrobial resistance (AMR) challenges in the past few years. This chapter explores the applications of AI in veterinary medicine, highlighting how it can empower veterinarians in their fight against AMR. AI-powered tool utilization is improving the management of antimicrobial agents while preserving the efficacy of antibiotics. These tools include predictive analytics, treatment optimization, drug development, surveillance systems, and educational programs. We also look at the challenges and possibilities of integrating AI into veterinary medicine, such as algorithm bias, data privacy, regulatory issues, and the requirement for multidisciplinary integration of technologies. Despite these challenges, AI technology can provide depth in studying and analyzing the issue more accurately and help implement more accurate measures to adopt antimicrobial resistance, giving veterinarians a powerful tool in their fight against AMR.

University of Petroleum and Energy Studies, Dehradun, India.

8.1 Introduction

Antimicrobial resistance (AMR) is a severe threat to the worldwide health and welfare of humans and animals. It is not just an environmental problem. Bacteria's ability to evade antimicrobial substances, a long-standing trait that has enabled their survival in an ever-changing environment, is now working against us. Prevention and treatment against infections in humans and animals have become substantially less effective due to the evolution of microorganisms in response to the selective pressure imposed by antibiotic treatments. This presents a serious threat to human health and well-being [1]. Antibiotics emerged in the modern era after Alexander Fleming discovered penicillin in 1928 [2], and after that, numerous people afflicted with diseases caused by pathogens have been saved by antibiotics. However, the extensive use of medicines in clinical practice has also led to drug resistance, and the situation is becoming increasingly urgent. Most common bacteria found in animal production systems include Escherichia coli [3–5], Enterococcus spp., and Staphylococcus aureus [6]. Foodborne zoonotic pathogens include non-typhoidal Salmonella (NTS) [4,7,8] and Campylobacter spp. [9]. The increasing prevalence of antibiotics in medical care and veterinary has resulted in drug resistance and elevated the risk of creating super-resistant bacteria, one of the most prevalent causes of antimicrobial resistance [10,11]. Poultry farming is one of the most popular types of farming in the food agriculture industry, as chicken is the most common meat consumed all over the globe, with over 90 billion tons of chicken meat produced yearly [12]. Most nations produce poultry utilizing an extensive array of antimicrobial agents [13,14], predominantly orally, to encourage productivity and development in addition to illness prevention and treatment [15]. Many antibiotics are extremely important and crucial for treating humans [16]. Antibiotic-resistant microorganisms (AMR) in ecological and pathogenic microorganisms are likely to develop earlier if antibiotics are used inappropriately in animal husbandry. Antimicrobial residues in meat [17] and eggs [18] raise health risks for humans, and there is concern over the introduction of AMR in bacteria from poultry farming. It is clear that the adoption of modern technologies is not just a choice but a necessity for strategic planning to mitigate antimicrobial risk in animal science.

Artificial Intelligence (AI) is a form of intelligence that simulates cognitive processes like comprehension, and both humans and animals can address problems related to themselves [19]. (Research on artificial intelligence (AI) in animal studies relates to the ability of systems based on computers or computer-controlled robots to carry out tasks that need intelligent entities. Artificial intelligence is a system that represents fundamental computer-based potential, simulating brain functions, including acquiring knowledge and solving issues seen in humans and animals. In comparison to the impact of artificial intelligence (AI) in other scientific domains like programming, philosophy, mathematics, neuroscience, cognitive science, and cybernetics, Animal health, public and community health, communicable illnesses, and livestock productivity are only a few of the topics covered in the broad discipline of veterinary medicine [20]. Due to its extraordinary advancement, exciting opportunities exist in AI's development, use, and clinical utilization in human and animal healthcare. When it comes to human healthcare, artificial intelligence (AI) has proven useful in target delivery of medicine, which modifies treatment in reaction to variations in genes and other characteristics; pharmaceutical research and exploration, where t he technique DeepTox, an algorithm based on deep learning, evaluates the adverse effects of compounds containing therapeutic molecules; diagnosis using images and surgical image analysis using computer vision [21]. This technology helps predict the risk of AMR and provides an in-depth analysis of reports related to animal studies.

8.2 Role of AI in Clinical Practice of Animal Studies

Artificial intelligence development, application, and integration have made remarkable progress and provide opportunities to improve the animal healthcare industry. The arrival of AI in veterinary medicine has brought fresh possibilities for enhancing the well-being of animals and their health. Unfortunately, given the quick speed and vitality of research and commercial product developments, these exciting opportunities also provide unique problems, notably in comprehending, interpreting, and accepting such powerful new technologies. Radiomics is a novel subfield of precision medicine related to AI that applies advanced mathematical techniques to a quantitative approach to biological imaging. It implies that medical reports and images contain information about infection-targeted processes outside human analysis reports.

Radiomics is the quantitative data-generating method for extracting various features from images. Analyzing these data with advanced statistical software becomes essential to developing individual patient-specific medicine. Radiomics applications are found in many medical imaging modalities, such as ultrasound, CT, MRI, and PET. Machine Learning (ML), a basic area of artificial intelligence, has a significant influence by developing algorithms that use knowledge from past data to create precise predictions for future data [22]. AI in A nimal H ealth (AH) can target more complex topics of research easily in a more understandable manner, including host-pathogen interactions, precision-based treatments for humans and animals, and quantitative and predictive epidemiology. AI can improve disease detection, precise diagnosis with few errors, understanding complex biological connections, treatment proposals, risk assessments, and target-specific therapies [23].

AI-Based disease detection: To implement efficient treatment to enhance patient outcomes and prognosis, early disease detection is crucial. Given that AI analyzes massive medical datasets, including imaging and clinical data, using sophisticated algorithms and machine learning, technology plays an essential role in this environment. AI in veterinary medicine has become vital for solving complicated issues in host-pathogen interaction, precision medicine for humans and animals, and statistical and predictive epidemiology. AI-driven systems powered by continuous monitoring ensure that at-risk patients are continuously monitored, facilitating quicker identification of alterations in medically associated indicators related with pathogens. To resolve these challenges in statistics and epidemiology, targeted drug delivery for both animals and humans, along with host-pathogen interaction, veterinary medicine must utilize artificial intelligence (AI) approaches. Solutions based on artificial intelligence for continuous surveillance ensure that at-risk patients are observed and facilitate rapid identification of alteration in targeted health indicators and markets. This is directly related and helpful for contagious diseases because it allows continuous patient management and early intervention. On the other hand, software-like surface and texture observation has been used to match and compare different diseases, such as glial cell neoplasia, with inflammatory meningoencephalitis. This comparison task is very difficult and time-consuming for the large sample data sets, continuous diagnostic imaging analysis, and overlapping characteristics, which result in interpretation [24]. Modern technology of artificial intelligence is integrating radionics and has been used as software that can support decision-making models and include them in conventional medical treatments and disease management for improving and enhancing sensitivity and accuracy of results through advanced statistical analysis approaches of Radionics [21].

Animal disease monitoring through Artificial intelligence: This intelligent technology holds great potential for monitoring and tracking infectious zoonotic diseases. In particular, AI's potential utilization in veterinary research and medicine shows immense potential to address issues related to recently discovered zoonotic animal infections [25]. Integrating artificial intelligence (AI) techniques with conventional disease control techniques, including automated learning and specialized algorithm models, provides novel opportunities for comprehending, foreseeing, and minimizing the impact of infections caused by zoonotic organisms [26]. Multiple artificial intelligence estimated models are designed with precision. They can be accessible for discovering risk-associated factors and estimating an individual's likelihood of developing an illness, enabling rapid detection and remediation [27]. Viral host identification and host range sensitivity have been predicted using web servers and conventional AI algorithm development. Algorithms based on artificial intelligence (AI), like the embedding layer of a Natural L anguage P rocessing (NLP) model, are related to extrapolate characteristics from viral nucleotides and protein sequences to predict the host ranges of influenza viruses [28]. Viral Host Prediction with Deep Learning (VIDHOP) is a deep learning technique that uses solely the viral nucleotide sequences to predict the hosts of viruses [29].

AI-Based disease models: Machine learning (ML) is advantageous in animal health monitoring, especially when generating prediction-making models. In the case of farms with a higher risk of acquiring a certain and becoming infected, it can be identified through AI technology. This is derived from a list of suspected risk variables and historical case data. Statistical-based machine models are related to cases of diarrhea viruses and estimate the amount of current and modern medicine in Canada [30]. Becker et al. employ artificial intelligence to identify which species of bats might operate as beta coronavirus reservoir hosts. Wardeh et al. predict modern insights into probable disease reservoirs through ML. Machine learning has also been used to analyze genomic sequencing data. Addressing big, complicated, hidden patterns might be intimidating using alternative techniques; machine learning is an appropriate response [31]. Furthermore, the accessibility of sequencing data has improved over the past few decades, Mostly as a consequence of reduced expenses and an equivalent rise in throughput at international medical and veterinary establishments.

When taken together, machine learning (ML) and whole genome sequencing yields improved whole genome sequencing provenance [32], pathogenicity assessment [33], antibiotic resistance phenotype [34], and clinical outcome prediction [35]. Put simply, better knowledge of the host-pathogen relationship and the onset and progression of disease improves molecular monitoring systems for most infections.

AI-based treatment, analysis, and evaluation: Vital sign monitoring is a crucial initial assessment before animal disease diagnosis and treatment. Veterinarians may electronically track health indicators using real-time monitoring devices that use AI-based analytics. This allows them to monitor patients' status, spot potential problems early on, and tailor their care. Veterinarians can create specific therapies with AI-driven decision support systems, which consider the animal's genetic makeup, past medical records, reaction to pharmaceuticals, and risk of anaphylaxis. This improvement enhances drug selection, dosage calculations, and evaluation of various treatment modalities. AI systems incorporate data analysis and modeling to give vets scientific guidance and globally acceptable protocols for treatment. AI improves decision-making precision by presenting specific therapy substitutes for each patient and using data from scientific literature, similar instances, and treatment outcomes. Long-term monitoring of diseases is vital for evaluating therapeutic outcomes and is essential to diagnosis and prognosis assessment. Models based on machine learning in dairy farming have been researched, especially regarding the diagnosis of illnesses such as mastitis [36] and lameness [37] and the prediction of calving time [38]. AI evaluates each animal's genetic predispositions, age, species, and medical history to determine its overall health profile.

This information is used to generate targeted treatment plans that are catered to each patient's special needs and characteristics. Veterinarians can adjust dosages, consider potential adverse effects, and target treatment strategies based on patient characteristics and current data. This procedure optimizes pharmaceutical regimens by analyzing the animal's response to therapy. Artificial Intelligence (AI) has significantly altered the struggle against antimicrobial resistance (AMR). It explores patterns in antibiotic use, identifies people at high risk of contracting an antimicrobial resistance illness, monitors the emergence of resistant bacteria, and quickly spots outbreaks of resistant diseases [39]. AI can establish and transform animal studies by providing more accurate, effective, and customized care for animal patients at the assessment, treatment, and assessment stages of veterinary patient care. Ongoing research and interaction related with veterinary practitioners and computer based academics will continue to improve and develop AI's uses in veterinary practice.

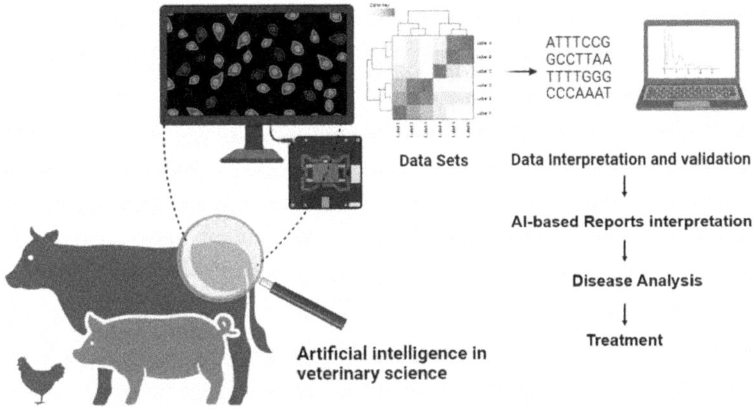

Figure 1: Artificial intelligence-based tolls evaluate, detect, and interpret the diseases associated with animals and provide targeted treatments.

8.3 The Role of Artificial Intelligence in Antimicrobial Resistance

The development of resistant microbes seriously threatens the worldwide health community, particularly in nations with weak healthcare systems. Therefore, tools that speed up the creation of new drugs, vaccines, and diagnostics are necessary. Due to the lengthy and challenging nature of current laboratory techniques, Artificial intelligence has an opportunity to significantly impact antimicrobial resistance and detect antibiotic residues in food products such as meat and dairy.

Consequently, because current testing techniques are labor-intensive and filled with difficulties, machine learning (ML) has the potential to significantly impact resistance to antibiotics and the identification of antibiotic residues in food, such as meat and milk.

AI can evaluate huge amounts of data to find therapeutic targets and create novel antibiotics with greater effectiveness and lower potential for resistance. Several AI-driven systems can examine surveillance data to monitor antibiotic resistance patterns, spot new dangers, and guide public health initiatives [40]. Particular artificial intelligence (AI) tools have been established for AMR research, including support vector algorithms over the design of new antibiotics, random forests for predicting the efficacy of antibiotic combination decision trees for determining the AMR burden and controlling the usage of antibiotics, and Naïve Bayes for the prediction of antibiotic resistance and identification of the primary factors of resistance [41]. Artificial Intelligence (AI) has been applied to the research and development of new approaches for whole genome sequencing and testing for antibiotic susceptibility. These novel techniques, which are quicker and more precise than the conventional approach, may enhance diagnoses and treatments for AMR [41]. The ideal stage of development for a new antibiotic is between ten and fifteen years, while the arrival of AI may enable an average time frame for development of only two years [42]. The implementation of acteria has completely transformed the battle against AMR, evaluated medication use sequences, and detected resistance infection outbreaks [39]. Humans become resistant to medicines when they eat antibiotic-contaminated meat, eggs, and dairy products from animals. Thus, antibiotic resistance will significantly reduce the utility of AI in livestock production in identifying and regulating the use of antibiotics in animals. Additionally, this tool will let scientists and researchers monitor the distribution of resistance genes. AI-based software can identify antibiotic residue in foods and track its origins, particularly in the dairy and meat industries. Therefore, AI is a promising technique or instrument that could significantly influence studies regarding AMR in humans and livestock. As AI expands, more tools to combat AMR will be created.

8.4 Discussion and Conclusion

AI is cutting the edge of addressing critical worldwide health concerns in various fields. To identify therapeutic targets, create novel antibiotics, and advise antibiotic usage while furthering the creation of crucial tools for the AMR war, AI provides a game-changing reaction to antimicrobial resistance (AMR). This is achieved by analyzing enormous datasets. The promise of artificial intelligence (AI) to enhance patient outcomes in cancer research can be observed in its incorporation into treatment development, treatments, and diagnostic methodologies [43]. Novel approaches are necessary due to the interconnectedness of zoonotic and re-emerging illnesses, and artificial intelligence (AI) in medical and veterinary research is emerging as a critical tool for early identification, containment, and precise diagnosis. AI has an important impact on pharmaceutical development and discovery since it may streamline procedures, find new therapeutic targets, and enhance clinical trial designs, all of which open up the possibility of more effective and efficient treatment options [44]. The method by which AI has been integrated into these various healthcare fields indicates how transformative an innovation it may be in changing the medical study and implementation field. The data sets enabling veterinary diagnostic imaging come with various fundamental challenges. First and foremost, managing the variety of species and breeds commonly seen in veterinary practice can be challenging, combined with the typical high patient caseload that necessitates collecting significant amounts of data. Second, a significant drawback is the lack of common cases of newly discovered illnesses for algorithm training; as a result, if such cases are not taken into account in advance, some diagnoses might not be identified. Animal veterinarians and other professionals should be encouraged to use veterinary databases like Veterinary Medical Databases (VMDB). Using current data formats and photos will operate as a basis for training AI algorithms for veterinary practice and research. Notwithstanding increased interest and efforts to use AI in human medical radiology [45], a significant knowledge gap remains before AI is fully adopted and successfully used in veterinary medicine. Combining new and conventional methods to develop a hybrid solution would be a feasible approach that might enable an easier transition to more potent and realistic predictive systems that are simple for both researchers and practitioners to implement [46–48].

References

[1] E. Palma, B. Tilocca, and P. Roncada. Antimicrobial resistance in veterinary medicine: An overview. International Journal of Molecular Sciences. 2020 Mar 11; 21(6): 1914.

[2] I. Ahmed, M. B. Rabbi, and S. Sultana. Antibiotic resistance in Bangladesh: A systematic review. International Journal of Infectious Diseases. 2019 Mar 1; 80: 54–61.

[3] D. A. Tadesse, S. Zhao, E. Tong, S. Ayers, A. Singh, M. J. Bartholomew, and P.F. McDermott. Antimicrobial drug resistance in Escherichia coli from humans and food animals, United States, 1950–2002. Emerging infectious diseases. 2012 May; 18(5): 741.

[4] G. A. Luna-Galaz, V. Morales-Erasto, C. G. Peñuelas-Rivas, P. J. Blackall, and E. Soriano-Vargas. Antimicrobial sensitivity of Avibacterium paragallinarum isolates from four Latin American countries. Avian diseases. 2016 Sep 1; 60(3): 673–6.

[5] C. Simoneit, E. Burow, B. A. Tenhagen, and A. Käsbohrer. Oral administration of antimicrobials increase antimicrobial resistance in E. coli from chicken–a systematic review. Preventive veterinary medicine. 2015 Jan 1; 118(1): 1–7.

[6] V. Bortolaia, C. Espinosa-Gongora, and L. J. Guardabassi. Human health risks associated with antimicrobial-resistant enterococci and Staphylococcus aureus on poultry meat. Clinical Microbiology and Infection. 2016 Feb 1; 22(2): 130–40.
[7] P. J. Blackall. Antimicrobial drug resistance and the occurrence of plasmids in Haemophilus paragallinarum. Avian Dis (1988) 32: 742–7. doi:10.2307/1590993.
[8] T. T. Van, H. N. Nguyen, P. M. Smooker, and P. J. Coloe. The antibiotic resistance characteristics of non-typhoidal Salmonella enterica isolated from food-producing animals, retail meat and humans in South East Asia. International Journal of Food Microbiology. 2012 Mar 15; 154(3): 98–106.
[9] C. H. Richter, B. Custer, J. A. Steele, B. A. Wilcox, and J. Xu. Intensified food production and correlated risks to human health in the Greater Mekong Subregion: a systematic review. Environmental Health. 2015 Dec; 14: 1–3.
[10] T. P. Van Boeckel, C. Brower, M. Gilbert, B. T. Grenfell, S. A. Levin, T. P. Robinson, A. Teillant, and R. Laxminarayan. Global trends in antimicrobial use in food animals. Proceedings of the National Academy of Sciences. 2015 May 5; 112(18): 5649–54.
[11] H. Bilal, M. N. Khan, T. Rehman, M. F. Hameed, and X. Yang. Antibiotic resistance in Pakistan: a systematic review of past decade. BMC infectious diseases. 2021 Dec; 21: 1–9.
[12] R. Twine. Emissions from animal agriculture—16.5% is the new minimum figure. Sustainability. 2021 Jun 2; 13(11): 6276.
[13] A. Agunos, D. Léger, and C. Carson. Review of antimicrobial therapy of selected bacterial diseases in broiler chickens in Canada. The Canadian Veterinary Journal. 2012 Dec; 53(12): 1289.
[14] M. F. Landoni, and G. Albarellos. The use of antimicrobial agents in broiler chickens. Vet J. (2015) 205: 21–7. doi:10.1016/j.tvjl.2015.04.016
[15] S. W. Page, and P. Gautier. Use of antimicrobial agents in livestock. Revue Scientifique et Technique-OIE. 2012 Apr 1; 31(1): 145.
[16] World Health Organization. Critically important antimicrobials for human medicine: ranking of antimicrobial agents for risk management of antimicrobial resistance due to non-human use. World Health Organization; 2017.
[17] M. Reig, and F. Toldrá. Veterinary drug residues in meat: Concerns and rapid methods for detection. Meat science. 2008 Jan 1; 78(1-2): 60–7.
[18] V. Goetting, K. A. Lee, and L. A. Tell. Pharmacokinetics of veterinary drugs in laying hens and residues in eggs: a review of the literature. Journal of veterinary pharmacology and therapeutics. 2011 Dec; 34(6): 521–56.
[19] S. Kour, R. Agrawal, N. Sharma, A. Tikoo, N. Pande, and A. Sawhney. Artificial intelligence and its application in animal disease diagnosis.
[20] E. Kumar. Artificial intelligence. IK International Pvt Ltd; 2013 Dec 30.
[21] A. Bohr, and K. Memarzadeh. The rise of artificial intelligence in healthcare applications. InArtificial Intelligence in healthcare 2020 Jan 1 (pp. 25–60). Academic Press.
[22] O. Bouhali, H. Bensmail, A. Sheharyar, F. David, and J. P. Johnson. A review of radiomics and artificial intelligence and their application in veterinary diagnostic imaging. Veterinary Sciences. 2022 Nov 8; 9(11): 620.
[23] P. Ezanno, S. Picault, G. Beaunée, X. Bailly, F. Muñoz, R. Duboz, H. Monod, and J. F. Guégan. Research perspectives on animal health in the era of artificial intelligence. Veterinary research. 2021 Dec; 52: 1–5.

[24] M. W. Wanamaker, K. M. Vernau, S. L. Taylor, D. D. Cissell, Y. G. Abdelhafez, and A.L. Zwingenberger. Classification of neoplastic and inflammatory brain disease using MRI texture analysis in 119 dogs. Veterinary Radiology & Ultrasound. 2021 Jul; 62(4): 445–54.

[25] S. Agrebi, and A. Larbi. Use of artificial intelligence in infectious diseases. InArtificial intelligence in precision health 2020 Jan 1 (pp. 415–438). Academic Press.

[26] W. Guo, C. Lv, M. Guo, Q. Zhao, X. Yin, and L. Zhang. Innovative applications of artificial intelligence in zoonotic disease management. Science in One Health. 2023 Nov 3: 100045.

[27] A. M. Sebastian, and D. Peter. Artificial intelligence in cancer research: trends, challenges and future directions. Life. 2022 Nov 28; 12(12): 1991.

[28] B. Xu, Z. Tan, K. Li, T. Jiang, and Y. Peng. Predicting the host of influenza viruses based on the word vector. PeerJ. 2017 Jul 18; 5: e3579.

[29] S. A. Babayan, R. J. Orton, and D. G. Streicker. Predicting reservoir hosts and arthropod vectors from evolutionary signatures in RNA virus genomes. Science. 2018 Nov 2; 362(6414): 577–80.

[30] T. Ajayi, R. Dara, and Z. Poljak. Forecasting herd-level porcine epidemic diarrhea (PED) frequency trends in Ontario (Canada). Preventive veterinary medicine. 2019 Mar 1; 164: 15–22.

[31] M. Wardeh, M. S. Blagrove, K. J. Sharkey, and M. Baylis. Divide-and-conquer: machine-learning integrates mammalian and viral traits with network features to predict virus-mammal associations. Nature Communications. 2021 Jun 25; 12(1): 3954.

[32] N. Munck, P. M. Njage, P. Leekitcharoenphon, E. Litrup, and T. Hald. Application of whole-genome sequences and machine learning in source attribution of Salmonella Typhimurium. Risk Analysis. 2020 Sep; 40(9): 1693–705.

[33] C. Deneke, R. Rentzsch, and B. Y. Renard. PaPrBaG: A machine learning approach for the detection of novel pathogens from NGS data. Scientific reports. 2017 Jan 4; 7(1): 39194.

[34] J. Sunuwar, and R. K. Azad. A machine learning framework to predict antibiotic resistance traits and yet unknown genes underlying resistance to specific antibiotics in bacterial strains. Briefings in Bioinformatics. 2021 Nov; 22(6): bbab179.

[35] P.M. Njage, P. Leekitcharoenphon, and T. Hald. Improving hazard characterization in microbial risk assessment using next generation sequencing data and machine learning: predicting clinical outcomes in shigatoxigenic Escherichia coli. International journal of food microbiology. 2019 Mar 2; 292: 72–82.

[36] A. S. Dhoble, K. T. Ryan, P. Lahiri, M. Chen, X. Pang, F. C. Cardoso, and K. D. Bhalerao. Cytometric fingerprinting and machine learning (CFML): A novel label-free, objective method for routine mastitis screening. Computers and electronics in agriculture. 2019 Jul 1; 162: 505–13.

[37] M. Taneja, J. Byabazaire, N. Jalodia, A. Davy, C. Olariu, and P. Malone. Machine learning based fog computing assisted data-driven approach for early lameness detection in dairy cattle. Computers and Electronics in Agriculture. 2020 Apr 1; 171: 105286.

[38] A.S. Keceli, C. Catal, A. Kaya, and B. Tekinerdogan. Development of a recurrent neural networks-based calving prediction model using activity and behavioral data. Computers and Electronics in Agriculture. 2020 Mar 1; 170: 105285.

[39] A. A. Rabaan, S. Alhumaid, A. A. Mutair, M. Garout, Y. Abulhamayel, M. A. Halwani, J. H. Alestad, A. A. Bshabshe, T. Sulaiman, M. K. AlFonaisan, and T. Almusawi. Application of artificial intelligence in combating high antimicrobial resistance rates. Antibiotics. 2022 Jun 8; 11(6): 784.

[40] T. Ali, S. Ahmed, and M. Aslam. Artificial intelligence for antimicrobial resistance prediction: challenges and opportunities towards practical implementation. Antibiotics. 2023 Mar 6; 12(3): 523.

[41] J. Lv, S. Deng, and L. Zhang. A review of artificial intelligence applications for antimicrobial resistance. Biosafety and Health. 2021 Feb 1; 3(01): 22–31.

[42] R. I. Masud, N. A. Fahim, M. L. Rana, M. S. Islam, and M. Tanvir. Artificial intelligence, a powerful tool to combat antimicrobial resistance: an update. *J Adv Biotechnol Exp Ther.* 2023; 6: 711–27.

[43] O. Elemento, C. Leslie, J. Lundin, and G. Tourassi. Artificial intelligence in cancer research, diagnosis and therapy. Nature Reviews Cancer. 2021 Dec; 21(12): 747–52.

[44] S. He, L. G. Leanse, and Y. Feng. Artificial intelligence and machine learning assisted drug delivery for effective treatment of infectious diseases. Advanced Drug Delivery Reviews. 2021 Nov 1; 178: 113922.

[45] M. M. Abuzaid, W. Elshami, H. Tekin, and B. Issa. Assessment of the willingness of radiologists and radiographers to accept the integration of artificial intelligence into radiology practice. Academic Radiology. 2022 Jan 1; 29(1): 87–94.

[46] O. C. Akinsulie, S. Shahzad, S. C. Ogunleye, I. P. Oladapo, M. Joshi, C. E. Ugwu, J.O. Gbadegoye, F. O. Hassan, R. Adeleke, Q. Afolabi Akande, and R. O. Adesola. Crosstalk between hypoxic cellular micro-environment and the immune system: a potential therapeutic target for infectious diseases. Frontiers in Immunology. 2023 Jul 31; 14: 1224102.

[47] K. Athanasopoulou, G. N. Daneva, P. G. Adamopoulos, and A. Scorilas. Artificial intelligence: the milestone in modern biomedical research. BioMedInformatics. 2022 Dec 17; 2(4): 727–44.

[48] R. Voelker. Diagnosing fractures with AI. Jama. 2018 Jul 3; 320(1): 23.

9

Ethical and Regulatory Considerations for Artificial Intelligence in Managing AMR

Chanchal, Ruby Dhiman, Prashant and *Ramendra Pati Pandey**

Concerns regarding the ethical and legal aspects of employing artificial intelligence (AI) in the management of antimicrobial resistance (AMR) are covered in this chapter. It draws attention to the necessity of transparent, equitable, and flexible policy frameworks for AI-driven programs. With a focus on the necessity of interdisciplinary cooperation and specifically created legislative frameworks, the chapter makes recommendations for dealing with these issues. Artificial Intelligence (AI) may successfully stop the spread of antimicrobial resistance (AMR) while maintaining moral principles and guaranteeing patient safety by assessing the moral and policy frameworks of the present.

9.1 Introduction

A natural phenomenon known as Antimicrobial Resistance (AMR) occurs when bacteria, viruses, fungi, and parasites learn to withstand the effects of medications meant to destroy them. AMR has emerged and spread more quickly as a result of the overuse and misuse of antibiotics in

School of health science and technology (SOHST), UPES, Dehradun.
* Corresponding author: ramendra.pandey@gmail.com

human health, animal husbandry, and the environment. According to the World Health Organization's (WHO) Global Antimicrobial Resistance and Use Surveillance System, AMR is becoming more common and is currently one of the main causes of death worldwide. In medicine, Artificial Intelligence (AI) has enabled the discovery of new medications and sped up the entire process of drug development and clinical research [1]. AI technology is an effective weapon in the fight against AMR.

In 1956, artificial intelligence (AI) was first mentioned in relation to technology that mimics human behavio r. Since then, there have been notable advancements in the discipline.

The term AI primarily refers to software or equipment that carry out tasks that are typically associated with intelligent beings. A recommendation by the OECD Council on Artificial Intelligence provides a precise definition of AI as follows: "An artificial intelligence system is a machine-based system that can make predictions, recommendations, or decisions that affect real or virtual environments for a given set of human-defined objectives." [2]. AI encompasses a range of methodologies, including N atural L anguage P rocessing (NLP), Deep Learning (DL), and M achine Learning (ML). Large Language Models (LLMs) are a class of artificial intelligence (AI) algorithms that comprehend, summarize, produce, and forecast new text-based material using deep learning methods and extraordinarily large data sets. Artificial intelligence (AI) has been used to examine massive datasets and find patterns that would be hard for humans to find. This has produced breakthroughs in areas like drug discovery and genomics [3]. Machine learning is used in most AI applications in healthcare, where computers pick up human tasks without explicit programming. Deep learning is a well-liked area of machine learning that uses intricate, multi-layered algorithmic frameworks and is particularly well-suited for input like speech or images [4]. The delivery, management, and optimization of healthcare services are all being completely transformed by the integration of AI. It has the ability to improve clinical judgment, expedite administrative procedures, and make it easier to create individualized treatment regimens that are catered to the requirements of each patient. Additionally, resource limitations, a growing patient population, and an abundance of health care information present opportunities that AI-driven solutions promise to solve, improving patient outcomes and building a more resilient healthcare ecosystem [5].

Rapid advancements in AI have given rise to instruments and applications in healthcare that frequently do not have governmental authorization, raising moral and legal questions. Consequently, in order to guarantee responsible development and useful use, it is essential to thoroughly investigate and comprehend the ethical and legal issues related to AI technologies in healthcare [2]. Transparency and traceability in the healthcare industry must adhere to even stricter regulations, taking a lead from other domains like public administration [6]. Legally speaking, data must abide by all laws, rules, and other legal obligations in all aspects of its gathering, storage, transmission, processing, and analysis. Furthermore, the law must continuously evolve to reflect the rapidly advancing state of technology, as well as how it is interpreted and applied [7].

However, in order to fully utilize AI in the fight against antimicrobial resistance (AMR), cooperation between researchers, politicians, healthcare practitioners, and technology developers is needed to get over ethical, legal, and technical obstacles.

9.2 Global Ethical Consideration in AI for AMR Management

A foundation of basic principles underpins medicine's ethical framework, directing healthcare providers in providing patient-centered, compassionate treatment. Respect for patient autonomy, which recognizes people's right to make educated healthcare decisions, is fundamental to these ideas. In keeping with the non-maleficence concept, the principle of beneficence emphasizes the healthcare provider's obligation to behave in the best interests of their patients, fostering wellbeing and attempting to maximize benefits while limiting damage [8].

Safeguarding patient privacy, traceability, accountability, and security must all be taken into account while developing digital healthcare systems. Protocols for handling any possible security breaches should also be established.

The WHO started creating a framework in 2019 to make it easier to incorporate technological advancements and digital innovations into healthcare. It is crucial to assess these technologies using criteria including "benefits, potential drawbacks, acceptability, feasibility, resource utilization, and considerations of equity," according to the WHO's recommendations for digital interventions in healthcare. These suggestions emphasize that these digital technologies have to be seen as vital tools in the pursuit of long-term sustainability and universal health care [9].

The ethical guidelines for using AI in AMR are intended to help developers, users, and regulators improve the layout and use of these technologies while guaranteeing appropriate supervision.

A complete collection of indications, recommendations, and guidelines for the development, implementation, and exploitation of AI technologies in medicine was created in June 2021 as part of a WHO project [10].

Figure 1 gives a brief synopsis of suggested ethical principles and guidelines for AI in healthcare.

Preservation of Liberty: Even in fields like the control of antibiotic resistance (AMR), the ethical precept of liberty preservation is essential to AI applications. Using AI in this way requires making sure that the choices AI systems make respect and preserve people's rights, liberty, and freedoms.

It entails taking precautions against probable biases or unfair treatment. To reduce the possibility of sustaining gaps in healthcare access and treatment, AI systems must be trained on a variety of representative datasets.

Moreover, collaboration amongst interdisciplinary stakeholders—technologists, legislators, ethicists, and civil society representatives—is essential to developing legal frameworks and guidelines that uphold the ethical and responsible development of AI while protecting individual

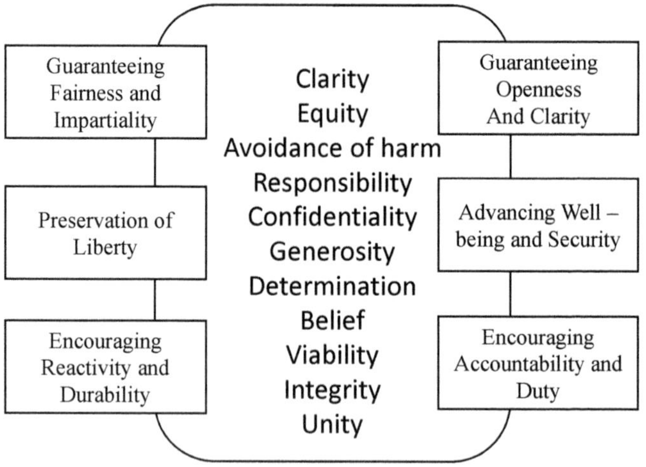

Figure 1: Ethical Principles and Guidelines.

liberty. While preserving the concept of liberty preservation, AI may be utilized to advance societal well-being, empower individuals, and open the door to a more democratic and equitable future.

***Advancing Well-being and Security*:** AI applications in AMR management must prioritize equality, safety, and human well-being, according to the ethical concept of enhancing well-being and security. This entails preventing unforeseen outcomes including a worsening of health disparities, protecting patient privacy, and preserving openness in AI algorithms and decision-making procedures.

The ethical idea of increasing security and well-being states that equity, safety, and human well-being should be given top priority in AI applications for AMR management. This means safeguarding patient privacy, averting unintended consequences like a worsening of health inequities, and maintaining transparency in AI algorithms and decision-making processes.

***Guaranteeing Openness and Clarity*:** Maintaining transparency and comprehensibility in AI systems is crucial when it comes to fighting antimicrobial resistance (AMR), since decisions made in this domain can have a direct influence on public health and treatment efficacy.

Explainability is the capacity of artificial intelligence (AI) systems to provide human-understandable explanations for their judgments. The openness of AI algorithms and their results, on the other hand, should be clear and free of any biases or ambiguities.

A crucial factor to take into account is finding a balance between improved accuracy and full algorithmic clarity (even at the expense of some accuracy).

***Encouraging Accountability and Duty*:** In the setting of antimicrobial resistance (AMR), where antibiotic abuse or overuse can worsen the issue, artificial intelligence (AI) can be extremely helpful in directing medical practitioners and policymakers toward more prudent antibiotic usage. But it's crucial that individuals in charge of developing and executing AI solutions give ethical issues related to responsibility and accountability top priority.

But as AI algorithms get more sophisticated and potent, it's critical to make sure they're responsible for their actions for the duration of their lives. Accountability ensures that AI systems function openly, equitably, and in accordance with social norms. It is both a moral obligation and a technological need.

It is the responsibility of healthcare practitioners to assess AI-driven AMR management solutions rigorously and ensure that they are clinically and evidence-based. They also need to push for patient-centered treatment methods and be on the lookout for any biases or unforeseen repercussions of AI algorithms.

This concept places a strong emphasis on being open, truthful, and dedicated to the greater good.

Guaranteeing Fairness and Impartiality: Fairness is one of the core principles of artificial intelligence (AI) ethics [11], and attempts to promote AI fairness have increased recently, as seen by the rise of publications emphasizing the need for change [12]. It is not only a moral goal but also a vital need to guarantee justice and equity in the use of AI to control antibiotic resistance (AMR). Populations across the world are impacted by AMR, and proper management requires equal access to prevention, diagnosis, and treatment.

Making sure AI algorithms don't promote biases or prejudice against specific groups is essential to guaranteeing justice. AI-powered diagnostic instruments, for example, need to deliver accurate findings across a range of demographics while avoiding biases that might disproportionately impact underrepresented populations.

Ensuring equitable distribution of AI-driven solutions for AMR control among all populations, irrespective of their socioeconomic level or geographic location, is a crucial aspect of equity. This might entail creating AI systems that are accessible, scalable, and flexible in environments with minimal resources.

In addition, continuous observation and assessment are required to recognize and resolve any unexpected outcomes or discrepancies that could occur.

Encouraging Reactivity and Durability: Encouraging responsiveness in the context of AI-driven AMR solutions means reacting quickly to new threats and changing antimicrobial resistance patterns. Large-scale real-time data analysis is made possible by AI algorithms, which also allow for the early discovery of emerging resistant strains, epidemic alarms, and resistance patterns. Healthcare systems can slow the emergence of resistance by implementing targeted interventions, including updated prescribing recommendations or infection control measures, as soon as these issues are recognized.

Durability also highlights the necessity of strong and long-lasting AI interventions. In addition to addressing urgent issues, AI systems must eventually help with the long-term sustainable management of AMR. This entails creating AI models that adjust to shifting resistance patterns, new approaches to therapy, and developments in the field of medicine. In addition, equitable distribution of benefits among varied populations and healthcare settings should be a top priority for permanent AI solutions.

9.3 Regulatory Landscape

As AI becomes more widely used in healthcare, appropriate governance is required to solve issues with trust, ethics, and regulations [13,14].

It is imperative that national and international rules be put into place in order to maintain control over regulated sectors, particularly healthcare. These laws are necessary to maintain medical ethics while allowing for the ethical integration of AI-driven technologies in healthcare [15].

Additionally, new research emphasizes how crucial it is to regulate AI technology at the healthcare system level in order to guarantee patient safety, healthcare system responsibility, increase adoption, boost clinician trust, and provide significant healthcare advantages [16].

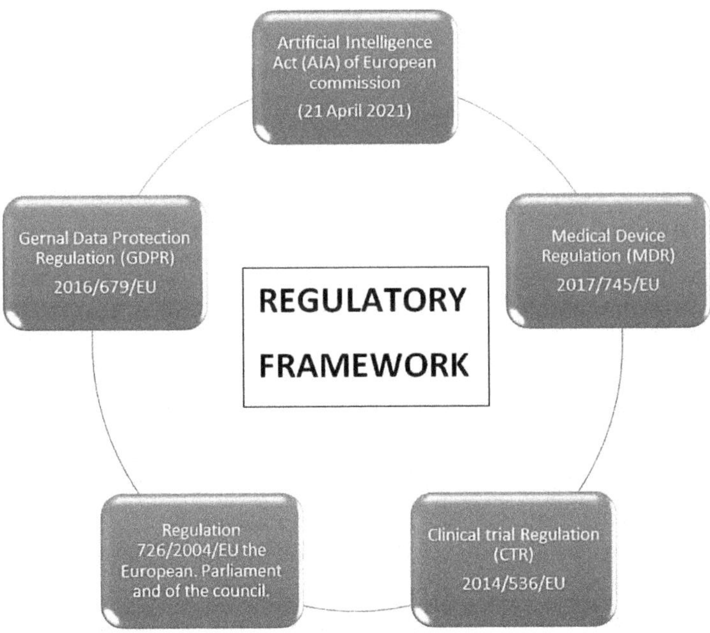

Figure 2: Key European Acts to govern AI in healthcare.

This section focuses on the creation and improvement of five crucial European acts that, when combined, can provide a single legislative framework for regulating artificial intelligence in healthcare (Fig. 2).

The General Data Protection Regulation (GDPR) was enacted by the European Union (EU) in 2018, marking the beginning of action in this attempt. GDPR was created expressly to safeguard personal data handled by controllers or data processors doing business in the EU. It also established a precedent for significant regulatory reforms in the US and Canada [17].

Additionally, the Artificial Intelligence Act (AIA), recently presented by the European Commission, aims to mitigate a number of hazards associated with the widespread application of AI [18,19]. This collection of rules aims to minimize or lessen possible risks associated with particular technological applications while promoting the responsible use of AI. This proposed legislation requires high-risk AI systems to go through post-market monitoring and pre-deployment compliance audits to make sure they are adhering to all of the AIA's standards [20].

In addition, the Medical Device Regulation (MDR) 2017/745/EU was just put into effect by the European Union [21], this takes the role of both the 93/42/EC Medical Device Directive (MDD) [22], as well as Regulation 90/385/EC, the Active Implantable Medical Device Directive (AIMDD) [23] which seeks to improve medical device industry regulation.

Generally speaking, the MDR's implementation leaves all obligations from the replacement regulatory acts in place. Rather, it presents fresh ones that highlight a life-cycle approach to device safety.

The following are the main modifications that the MDR's implementation and enforcement for medical devices have brought about [2]:

i) The MDR improves controls to guarantee the security and efficiency of gadgets.
ii) Not all medical devices may be used with the mechanism that expedites their introduction onto the market or into service by equating them with already-existing equipment.
iii) All medical devices are subject to post-marketing clinical follow-up, which has raised the significance and quantity of clinical assessments and investigations.

Medical equipment fall into one of four primary classifications, as stated under Article 51(1) of the MDR: I, IIa, IIb, and III. Based on the standards outlined in Annex VIII, this categorization is determined by the products' intended use and inherent dangers. The majority of non-invasive, passive devices, or class I devices, are often the lowest risk. Devices in class II(a) have a medium risk, while those in class II(b) carry a medium to high danger. Devices that pose a high danger are under Class III. Software is categorized as class II(a) when it is meant to give information for decisions about diagnosis or treatment, unless those decisions could lead to a patient's death or irreversible health decline (in which situation it is classified as class III) or to a serious decline in their health or the need for surgical treatment (in that case it is categorized as class II(b)). In addition, software intended for physiological process monitoring is categorized as Class II(a), unless it is intended for the monitoring of critical physiological parameters, where variations in these parameters may immediately endanger the patient (in which scenario it is categorized as Class IIb).

All other software is classified as class I (see Annex VIII, rule 11). It is important to remember that class I risk categorization is usually not relevant to AI-based medical devices. From this perspective, it should be taken into consideration that danger class I could not possibly apply in terms of AI-based MD [24]. However, according to the MDR's definition, artificial intelligence technologies are classified as medical devices. But further explanation is needed for this conclusion. In particular, whether a medicinal product integrated into an AI device is deemed "ancillary," subject to the MDR, or "non-accessory," subject to rules pertaining to pharmaceuticals for human use, determines which legislation applies [25]. In this case, one may contend that the medical product integrated into the device falls under the MDR's review and authorization procedure if it serves an auxiliary function for the device, as stated in Article 1 of the MDR. The comprehensive product, however, will fall under the purview of either Regulation 726/2004/EC, which controls Community procedures for authorizing and supervising medicinal products for human and veterinary use, including the creation of the European Medicine Agency (EMA), when applicable, or Directive 2001/83/EC of the European Parliament and of the Council, dealing with medicinal products for human use.

Furthermore, the Clinical Trial regulation (CTR) provides pertinent recommendations for legislation when it comes to accountability and openness in AI.

A number of obligations are outlined in Article 29 AIA for users who utilize high-risk AI systems. These duties include following the usage guidelines that have been provided, making sure that the data entered is relevant to the system's intended use, keeping a close eye on how the system is operating in accordance with the guidelines, notifying the innovator or distributor right away, and stopping the use of the AI medical equipment immediately if there are any concerns that its application is not adhering to the guidelines. Additionally, users must use pertinent information to fulfill their obligations, report serious hazards to the sponsor and investigator, and abide by the data protection guidelines outlined in Articles 27 of Directive (EU) 2016/680 or Article 35 of Regulation (EU) 2016/679, as applicable.

Safeguarding that AI honors human rights, democratic values, and the rule of law is emphasized in both the UNESCO Ethics of AI Recommendation and the OECD Guidelines for AI [26].

9.4 Data Privacy and Security

Modern security relies heavily on data protection, and encryption methods are important for this. Their main objective is to transform data into a format that cannot be decoded without a special key [27]. Encryption techniques use mathematical processes to convert plaintext data into ciphertext, which appears to be gibberish to anybody lacking the necessary decryption key [28,29]. Among its many useful applications are the protection of sensitive data in databases, the encryption of online conversations, and the preservation of files on electronic devices [30], (see Fig. 3 and 4).

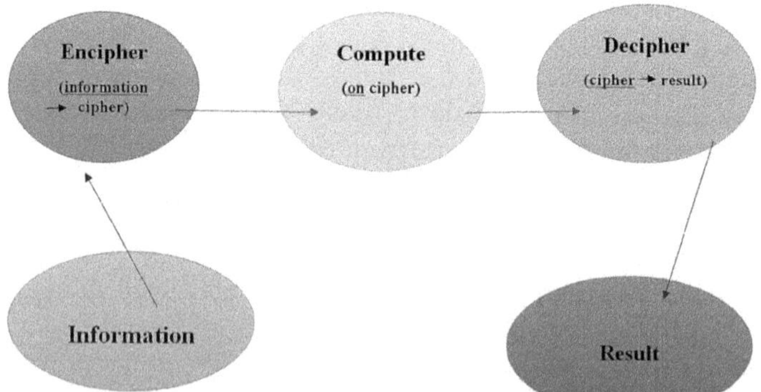

Figure 3: Steps involved in Homomorphic Encryption.

Figure 4: Understanding the notion of differentiated privacy.

Homomorphic encryption protects encrypted data by allowing calculations on it. First, the data is encrypted; next, calculations are made; and last, the results are decrypted.

A type of encryption known as homomorphic encryption makes it possible to compute encrypted data without having to access the plaintext. With this method, all operations are performed on the encrypted data, and the important information is only revealed when the result is decrypted. Numerous uses exist for homomorphic encryption, such as safe cloud computing, encrypted search, and calculations in medical research that protect anonymity [27].

By guaranteeing that the inclusion or removal of one point of information has no appreciable effect on the findings of any study, the technique known as Differential Privacy safeguards individual data points. This is accomplished by adding noise to the information or running queries over the data, which obscures specific items. Differential privacy is helpful for safeguarding user data in census databases, retaining security in machine learning datasets, and keeping user information in analytics [32].

A method called Secure Multi-Party Computation (SMPC) allows a function to be privately computed over the inputs of several participants. The information is split up and distributed to other parties. These discrete elements are processed using calculations that conceal the source data. Safe voting practices, cooperative data-driven initiatives without sharing information, and privacy-preserving medical research are just a few of the numerous uses for SMPC.

To sum up, encryption methods such as SMPC, Differential Privacy, and Homomorphic Encryption are essential to contemporary data security. They offer a strong and safe method for maintaining privacy, safeguarding sensitive data, and enabling safe calculations on encrypted data [32, 33].

9.5 Addressing Ethical and Regulatory Challenges

There are considerable risks and expenses involved in transforming AI and massive amounts of data into safe and effective real-world healthcare apps, services, and processes. As a result, protecting the business goals of AI and data-driven healthcare systems has become a more important topic [34]. In the past, vital indicators including blood pressure, glucose levels, and heart rate could only be measured by medical specialists [35]. However, the ongoing gathering of this kind of data is now possible thanks to modern mobile applications.

However, it's critical to address the ethical concerns raised by the use of AI, especially with regard to patient autonomy, informed permission, and data confidentiality and privacy violations [34,35]. Strong data protection laws are essential to protecting personal privacy in light of the growing role that big data and artificial intelligence (AI) are playing in healthcare and precision medicine. Nations all around the world have passed legislation to safeguard people' privacy, such as the General Data Protection Regulation (GDPR) in Europe and the Health Insurance Portability and Accountability Act (HIPAA) in the United States [36, 37]. While the GDPR has established substantial data protection regulation inside the EU, causing a dramatic worldwide change in data protection, HIPAA primarily covers pertinent health information supplied by covered businesses [36,37].

Increased cyberattacks are a key factor that can compromise information about patients, interfere with vital healthcare processes, and endanger patient safety when AI is used in the healthcare system [38, 39]. These cyberthreats can be identified and avoided with the use of predictive algorithms. Conducting a thorough investigation into cybersecurity and the cyber risk environment of healthcare systems is crucial in order to protect patient privacy and preserve system integrity [38,40]. The danger of depending just on one effective AI algorithm can be reduced by putting in place a number of strong AI algorithms [38,40]. While security lapses and data privacy are issues with AI within healthcare, it has several benefits, including the ability to streamline tasks, increase productivity, save time and resources, support for research, and lessen stress for doctors [39]. When assessing the effects of digital technology on participants in the healthcare supply chain, a theoretical framework for ethical evaluation has been presented that places a priority on ethical awareness, openness, and responsibility [40,41].

However, several characteristics of AI technology complicate ideas of responsibility. These characteristics include their software complexity, opacity, dependence on human input, interactions, discretionary functions, scalability, and ability to reveal hidden insights. A specific obstacle in assigning accountability stems from the "control problem" linked to artificial intelligence. Since AI-driven systems are capable of operating autonomously and evolving in ways that the creator may claim are unanticipated, under this situation, AI creators and designers may not be held liable [42]. While expectations of manufacturers of other widely used medical technologies, such as drug and vaccine manufacturers, medical device companies, and healthcare equipment manufacturers, are already well established, assigning responsibility to the developer could provide a reason to take all reasonable steps to minimize harm to the patient.

The "traceability of harm" is another difficulty. This problem persists even in the absence of artificial intelligence in complicated decision-making systems in the healthcare and other sectors. As a result of the several agents involved in the creation of AI, assigning accountability is a difficult undertaking that involves ethical and legal considerations. When developers and users are not held accountable, it can lead to a lack of compensation for those who have been harmed, an incomplete understanding of the harm and its underlying causes, ignored harm, and a possible decline in public trust in these technologies. These are just a few of the negative effects of responsibility diffusion [43].

However, it is unlikely that AI in the medical domain would achieve total autonomy; rather, it may advance to a point of conditional automation or require continuous human assistance [44].

The use of AI in medicine threatens the basic precepts of informed consent and the public's general faith in healthcare, particularly when its usage is concealed. There is a serious risk associated with giving AI technology the power to make decisions about patients, even though some people may see this as a method to increase patient autonomy. This danger increases if the patient believes that the technology provides a better insight of their health state and prognosis than a doctor does [45].

9.6 Stakeholder Engagement and Collaboration

Involving stakeholders is essential to the development and application of artificial intelligence (AI) technology. It is essential to make sure AI technologies are developed and applied in a way that is morally righteous, conscientious, and advantageous to society. AI systems have a significant influence on many facets of human life, including banking, healthcare, education, transportation, and security. To comprehend and resolve the ethical issues raised by new technologies, it is crucial that developers, users, and impacted communities collaborate and communicate with one another [27].

Healthcare stakeholders should create and implement a well-thought-out plan for using artificial intelligence (AI) in healthcare to manage the associated costs, technological infrastructure, and integration of AI systems into clinical processes. Additionally, because of the unadvertised hazards and unreliability of AI-based clinical decision support systems, doctors often lack knowledge and mistrust these systems [46]. Its skepticism presents a significant barrier to widespread acceptance. Implementing understandable AI solutions is becoming more and more important in order to overcome this, increase user confidence, and deal with these problems [47].

In order to successfully include stakeholders, it is imperative to establish channels and procedures that facilitate transparent dialogue and cooperative efforts. Public consultations, stakeholder seminars, and cross-sector collaboration are a few examples of these venues. These forums need to uncover issues, collaboratively develop solutions, foster trust, and advance literacy [27].

For a system to be implemented successfully, professional stakeholders must come to believe in its effectiveness. In this sense, classical AI models provide a unique difficulty. The "black box" problem is the outcome of their non-deterministic and correlational character, which is different from causal: The decision-making mechanism of the system is opaque to the user [48,49]. Though trust is a fundamental need for AI in healthcare, there is a dearth of research on the subject. Regarding what trust is and how to build it in the healthcare industry, Gillie et al. [50] could not find agreement.

As a prerequisite for reliable AI, few studies primarily concentrate on the explainability and interpretability of algorithms [51,52,53]. In order to simplify things, trust must be regarded as a continuous process of building faith between the persons concerned [54,55]. This may be accomplished by putting in place transparent standards, ethical codes of behavior, and explicit accountability systems. Increasing the level of AI literacy among all parties involved might facilitate more fruitful and knowledgeable dialogues on the moral implications of AI. This is known as literacy promotion. This entails teaching interested parties about the advantages and disadvantages of artificial intelligence (AI), the technological constraints of AI, and the moral precepts that drive AI research and use. Emotional ties are crucial for interpersonal trust; artificial intelligence (AI) shouldn't require emotional trustworthiness in order to be reliable; instead, it should represent reliability on par with other technologies that assist in medical decision-making [56].

9.7 Future Directions and Recommendations

AI has the potential to completely transform healthcare, but in order to do so, a number of issues need to be resolved. One of these issues is the dearth of reliable medical data, which might produce unreliable results. The use of AI in clinical practice may also be constrained by issues with data security, availability, and privacy. Achieving the intended results also depends on using the right approach and identifying pertinent clinical measures. Artificial intelligence (AI) has the potential to magnify human bias in the creation and use of its tools if it is not properly supervised [57]. This can be especially troublesome in delicate areas like patient care. Therefore, the creation of AI technologies has consequences for the education of health professionals today, emphasizing the need to acknowledge human fallibility in domains such as clinical reasoning and evidence-based medicine [58].

It will need a multidisciplinary approach, creative data annotation techniques, and the creation of more rigorous AI models and approaches to address these issues and offer useful answers. The AI community will have the chance to design an integrated best practice framework for implementation and maintenance by combining current best practices for ethical inclusion, software development, implementation science, and human-computer interaction [59]. Furthermore, for AI to be successful in clinical practice, cooperation between different medical settings is needed to share data, guarantee its quality, and validate the results of analyses. To ensure proper adaptation to deliver the best patient care and avoid any legal or ethical issues or misunderstanding of the results without verifying the results, a further recommendation is to provide appropriate training and education that begins at the undergraduate level for all health care practitioners and continues with constant growth and enhancement for the practitioners working in the current practice [60].

According to a research done among radiology residents, 86% of students thought AI will transform and improve their field, and up to 71% said medical schools should teach AI to increase comprehension and application [61]. Through this integration, it is ensured that aspiring healthcare professionals are well-versed in AI and its applications from the outset of their studies.

More phases of drug development, such as molecular design, dosage regimen prediction and related effectiveness prediction, dynamic protein-protein interaction modeling [62], ligand-based simulation, quantitative structure-activity connection modeling, molecular representations and high-order feature extraction [63], drug repurposing, and toxicity reduction, are likely to involve artificial intelligence in the future. The present big data age (huge information from high-throughput technologies, genomic/expression databases) and the quick development of AI might hasten the generation of new chemical entities with unique MOAs. Antibiotics or antibacterial biologicals are therefore anticipated to pass the medication development process and reach the market in greater numbers. The delayed development of antibiotic resistance could greatly aid the battle against resistant bacteria, especially with the goal of combining new treatments with antibiotic stewardship [1].

The *in vitro* phases of the drug-discovery pipelines for innovative antibiotics have been considerably reduced using AI. However, because biologicals are complicated, artificial intelligence has not done much to help understand how such medications function in the human body. There is significantly less data available than what is needed to train AI algorithms to answer queries *in vivo* [64]. It is critical to find a balance between data accessibility and confidentiality in order to handle this issue. Federated learning, active learning, and swarm learning are a few interesting directions that merit more investigation [64,65].

It is impossible to emphasize the significance of high-quality wet experimental data for any AI-based antibacterial study [66]. The combination of precision medicine, biomarker identification, and disease progression with drug discovery will be a future issue. AI has the potential to assist in the development of systematic models that facilitate quick illness diagnosis and automatically suggest more individualized patient therapies [67,68], thanks to developments in single-cell and high-throughput technologies. Additionally, it could inspire drug developers to shift from biomarker identification to medication development, transforming drug discovery and clinical practice in the future [69].

AMR, which results from the improper usage of antibiotics, is a complicated problem that goes beyond the creation of novel medications and treatments. A cooperative and comprehensive teaching strategy for the public and healthcare professionals about the appropriate use of antibiotics is also necessary to solve this issue [1].

9.8 Conclusion

In conclusion, the development of antimicrobials has been sluggish even though antimicrobial resistance (AMR) poses a severe danger to human health and has received a lot of attention. Between 2014 and 2021, just 18 new antibiotics were authorized, and only two of those had a novel MOA.

AI's application in healthcare has the potential to completely transform patient outcomes and treatment. Predictive analytics powered by AI can improve clinical laboratory testing and illness detection in terms of precision, efficacy, and cost-effectiveness. AI may also help with managing population health and the creation of guidelines by optimizing pharmaceutical choices and delivering accurate, real-time information. Improving patient care has been demonstrated by the integration of AI in virtual medical and psychological assistance. To guarantee fair and efficient usage of AI, it is crucial to address issues like prejudice and lack of customisation.

In order to guarantee the ethical and efficient application of AI in healthcare, a number of steps must be followed. First and foremost, in order to safeguard patient data and vital healthcare activities, thorough cybersecurity plans and strong security measures need to be created and put into place. In order to create standards and recommendations for AI algorithms and their application in clinical decision-making, cooperation between regulatory agencies, healthcare organizations, and AI researchers is essential. Research & development expenditures are also required to progress AI solutions suited to healthcare issues [70].

AI systems are able to continually analyse variables including geographic distribution, illness prevalence, and population demographics. This can help with therapy or prevention by identifying patients who are more likely to develop specific illnesses.

Trust between patients and providers, ethical standards, and strong AI systems all depend on stakeholder collaboration. To fully realize the promise of artificial intelligence in healthcare, more research, creativity, and multidisciplinary cooperation are required. AI is predicted to transform healthcare with effective integration, improving patient outcomes, efficiency, and access to individualized care and high-quality therapy.

The draft voluntary AI code of conduct's pillars depend on openness, risk assessment, and other technical information about AI system development. Accelerating the creation of the US-EU AI code of conduct would be a step toward harmonizing European and American AI laws and regulations. Because AI technologies are not country-specific, it is uncertain whether global regulatory convergence will ever take place, although it would be advantageous given the common healthcare concerns that exist worldwide.

AI can assist physicians in making better judgments, even if it might not be able to fully replace clinical judgment. AI might be used to do screening and assessment if there is a lack of medical expertise in an environment with limited resources. All AI decisions, even the shortest ones, are methodical since algorithms are involved, in contrast to human decision-making. Because of this, even in cases when activities don't have legal consequences (due to the lack of effective legal frameworks), they always result in accountability—not from the machine itself, but from the people who created it and the people who use it. Artificial intelligence (AI) in healthcare is expected to begin with limited AI use and either coexist or replace present systems, despite ethical dilemmas surrounding its application.

While there are ethical dilemmas with adopting AI, it is likely to supplement, coexist with, or replace present systems, ushering in the age of artificial intelligence in healthcare. It is also immoral and unscientific not to employ AI.

Lastly, it is important to maintain regulatory frameworks and ethical standards up to date with the rapidly changing use of AI in healthcare.

References

[1] G. Y. Liu, D. Yu, M.M. Fan, *et al.* Antimicrobial resistance crisis: could artificial intelligence be the solution?. *Military Med Res* **11**, 7 (2024). https: //doi.org/10.1186/s40779-024-00510-1.

[2] C. Mennella, *et al.* "Ethical and regulatory challenges of AI technologies in healthcare: A narrative review." *Heliyon* vol. 10,4 e26297. 15 Feb. 2024, doi:10.1016/j.heliyon.2024.e26297.

[3] S. A. Alowais, *et al.* "Revolutionizing healthcare: the role of artificial intelligence in clinical practice." *BMC medical education* vol. 23,1 689. 22 Sep. 2023, doi:10.1186/s12909-023-04698-z.
[4] R. Rokhshad, *et al.* "Ethical considerations on artificial intelligence in dentistry: A framework and checklist." *Journal of dentistry* vol. 135 (2023): 104593. doi:10.1016/j.jdent.2023.104593.
[5] S. Yelne, et al. "Harnessing the Power of AI: A Comprehensive Review of Its Impact and Challenges in Nursing Science and Healthcare." *Cureus* vol. 15,11 e49252. 22 Nov. 2023, doi:10.7759/cureus.49252.
[6] H. P. Olsen J. L. Slosser T. T. Hildebrandt, and C. Wiesener C. What's in the box? The legal requirement of explainability in computationally aided decision-making in public administration. SSRN Scholarly Paper. Rochester: Social Science Research Network; 2019. https: //doi.org/10.2139/ssrn.3402974.
[7] D. Schönberger. Artificial intelligence in healthcare: a critical analysis of the legal and ethical implications. *Int J Law Inf Technol.* 2019; 27: 171–203.
[8] T. L. Beauchamp, and J. F. Childress. Principles of Biomedical Ethics, 2001, Oxford University Press.
[9] W. Guideline Recommendations on Digital Interventions for Health System Strengthening World Health Organization (2019) 2020-10.
[10] S. J. Russell, and P. Norvig. *Artificial intelligence: a modern approach.* Pearson, 2016.
[11] B. D. Mittelstadt, P. Allo, M. Taddeo, S. Wachter, and L. Floridi. The ethics of algorithms: Mapping the debate. Big Data Soc. 2016; 3: 2053951716679679. SAGE Publications Ltd. 10.1177/2053951716679679
[12] P-H. Wong. Democratizing algorithmic fairness. *Philos Technol.* 2020; 33: 225–244. doi: 10.1007/s13347-019-00355-w.
[13] S. Reddy, et al. "A governance model for the application of AI in health care." *Journal of the American Medical Informatics Association: JAMIA* vol. 27,3 (2020): 491–497. doi:10.1093/jamia/ocz192.
[14] Guidance, W. H. O. "Ethics and governance of artificial intelligence for health." *World Health Organization* (2021).
[15] House of Lords Report. AI in the UK: Ready, Willing and Able? HOUSE OF LORDS Select Committee on Artificial Intelligence Report of Session 2017–2019, p–36, 2018.
[16] J. S. Marwaha, A. B. Landman, G. A. Brat. *et al.* Deploying digital health tools within large, complex health systems: key considerations for adoption and implementation. *npj Digit. Med.* **5**, 13 (2022). https: //doi.org/10.1038/s41746-022-00557-1.
[17] M. B. Forcier, H. Gallois, S. Mullan, and Y. Joly. Integrating artificial intelligence into health care through data access: can the GDPR act as a beacon for policymakers?, *Journal of Law and the Biosciences*, Volume 6, Issue 1, October 2019, Pages 317–335, https: //doi.org/10.1093/jlb/lsz013.
[18] European Commission. Commission Staff Working Document, Impact Assessment, Accompanying the Proposal for a Regulation Laying down Harmonized Rules on Artificial Intelligence (Artificial Intelligence Act), SWD (2021) 84 final, Part ½, (2021), p. 1.

[19] Artificial Intelligence (Artificial Intelligence Act)European Commission. Proposal for a Regulation Laying down Harmonised Rules on COM (2021) 206 final, European Commission, Brussels, Belgium (2021), p. 206.
[20] M. Shake. The European Commission's Artificial Intelligence At Stanford University Human-Centered Artificial Intelligence (BY), Stanford, Canada (2021), 2021-06.
[21] Regulation (EU) 2017/745 of the European Parliament and of the Council of 5 April 2017 on Medical Devices, Amending Directive 2001/83/EC, Regulation (EC) No 178/2002 and Regulation (EC) No 1223/2009 and Repealing Council Directives 90/385/EEC and 93/42/EEC.
[22] Council Directives 90/385/EEC and 93/42/EEC.
[23] I. Habli, et al. "Artificial intelligence in health care: accountability and safety." *Bulletin of the World Health Organization* vol. 98,4 (2020): 251–256. doi:10.2471/BLT.19.237487.
[24] N. Martelli, et al. "New European Regulation for Medical Devices: What Is Changing?." *Cardiovascular and interventional radiology* vol. 42,9 (2019): 1272–1278. doi:10.1007/s00270-019-02247-0.
[25] A. Migliore. "On the new regulation of medical devices in Europe." *Expert review of medical devices* vol. 14,12 (2017): 921-923. doi:10.1080/17434440.2017.1407648.
[26] OECD. "The Organization for Economic co-operation and Development (OECD)—United States Department of State." Accessed: Jul. 07, (2023).
[27] P. Radanliev, O. Santos, A. Brandon-Jones, and A. Joinson. Ethics and responsible AI deployment, Frontiers in Artificial Intelligence, Vol. 7(2024), doi:10.3389/frai.2024.1377011
[28] NIST. "Advanced encryption standard (AES)," (2001). Accessed: Mar. 19, 2023.
[29] J. Bergstra, R. Bardenet, Y. Bengio, B. Kégl, and R. Bardenet. "Algorithms for hyper-parameter optimization," Neural information processing systems foundation, (2011). Accessed: Mar. 16, 2021.
[30] B. Rishi, K. Kevin, Z. Daniel, and L. Percy. *Do Foundation model providers comply with the draft EU AI act?* Stanford: Center for Research on Foundation Models. Institute for Human-Centered Artificial Intelligence, 2023.
[31] C. S. Thirumalai, S. Budugutta, and C. Thirumalai. "Public key encryption for SAFE transfer of one time password heuristic prediction of olympic medals using machine learning view project extreme machine learning view project public key encryption for SAFE transfer of one time password", Accessed: Sep. 22, (2023).
[32] M. Zheng, D. Xu, L. Jiang, C. Gu, R. Tan, and P. Cheng. "Challenges of privacy-preserving machine learning in IoT," in *Proceedings of the First International Workshop on Challenges in Artificial Intelligence and Machine Learning for Internet of Things*, 1–7, 2019.
[33] A. Cossy-Gantner, S. Germann, N. R. Schwalbe, and B. Wahl.. Artificial intelligence (AI) and global health: how can AI contribute to health in resource-poor settings? *BMJ Glob. Health* 3,4: e000798. doi:10.1136/BMJGH-2018-000798, 2018.
[34] S. Gerke, T. Minssen, and G. Cohen. Ethical and legal challenges of artificial intelligence-driven healthcare. Artif Intell Healthc. 2020; 295–336. 10.1016/b978-0-12-818438-7.00012-5.

[35] M. Subramanian, A. Wojtusciszyn, L. Favre, S. Boughorbel, J. Shan, K. B. Letaief, et al. Precision medicine in the era of artificial intelligence: implications in chronic disease management. *J Transl Med.* 2020; **18**(1): 472. doi: 10.1186/s12967-020-02658-5.

[36] I. G. Cohen, and M. M. Mello. HIPAA and protecting health information in the 21st Century. *JAMA.* 2018; **320**(3): 231. doi: 10.1001/jama.2018.5630.

[37] B. Yuan, and J. Li. The policy effect of the General Data Protection Regulation (GDPR) on the digital public health sector in the European Union: an empirical investigation. *Int J Environ Res Public Health.* 2019; **16**(6): 1070. doi: 10.3390/ijerph16061070.

[38] S. Alotaibi, R. Mehmood, I. Katib, O. Rana, A. Albeshri, and A. Sehaa. Big Data Analytics Tool for Healthcare symptoms and Diseases Detection using Twitter, Apache Spark, and machine learning. *Appl Sci.* 2020; **10**: 1398. doi: 10.3390/app10041398.

[39] M. Abdel-Hameed Al-Mistarehi, M. Mijwil, F. Youssef, M. Bounabi; G. Ali, and M. Abotaleb. Artificial Intelligence Solutions for Health 4.0: Overcoming Challenges and Surveying Applications. MJAIH 2023, 2023, 15–20.

[40] P. Radanliev, D. De Rour. Epistemological and bibliometric analysis of Ethics and Shared responsibility—health policy and IoT Systems. *Sustainability.* 2021; **13**(15): 8355. doi: 10.3390/su13158355.

[41] P. Radanliev, D. De Roure, U. Ani, and G. Carvalho. The ethics of shared Covid-19 risks: an epistemological framework for ethical health technology assessment of risk in vaccine supply chain infrastructures. *Health Technol (Berl)* 2021; **11**(5): 1083–91. doi: 10.1007/s12553-021-00565-3.

[42] T. S. N. C. on Medical Ethics in brief –artificial intelligence in healthcare https://smer.se/wp-content/uploads/2020/06/smer-2020-2-in-brief-artificial-intelligence-in-healthcare.pdf (2020).

[43] K. Yeung. A study of the implications of advanced digital technologies (including ai systems) for the concept of responsibility within a human rights framework, MSI-AUT(2018)05, (2018).

[44] P. Humphreys. The philosophical novelty of computer simulation methods, Synthese, 169 (2009), pp. 615–626.

[45] J. Ross, and C. Webb, and F. Rahman. Artificial Intelligence in Healthcare, Academy of Medical Royal Colleges, London (2019).

[46] A. Choudhury, and O. Asan. Impact of accountability, training, and human factors on the use of artificial intelligence in healthcare: exploring the perceptions of healthcare practitioners in the US, Hum. Factors Healthc., 2 (2022), Article 100021.

[47] F. Giuste, *et al.* "Explainable Artificial Intelligence Methods in Combating Pandemics: A Systematic Review." *IEEE reviews in biomedical engineering* vol. 16 (2023): 5-21. doi:10.1109/RBME.2022.3185953.

[48] W. Samek, T. Wiegand, and K-R. Müller. Explainable artificial intelligence: understanding, visualizing and interpreting deep learning models. arXiv: 170808296, 2017.

[49] B. Sahiner, A. Pezeshk, L. M. Hadjiiski *et al.* Deep learning in medical imaging and radiation therapy. Med Phy 46: e1–e36, 2019.

[50] F. Gille, A. Jobin, and M. Ienca. What we talk about when we talk about trust: theory of trust for AI in healthcare. Intell-Based Med 1–2: 100001, 2020.

[51] J. D. Fuhrman, N. Gorre, Q. Hu, H. Li, I. El Naqa, M. L. Giger. A review of explainable and interpretable AI with applications in COVID-19 imaging. *Med Phys*, 49: 1–1, 2022.

[52] B. Heinrichs, and S. B. Eickhoff. Your evidence? Machine learning algorithms for medical diagnosis and prediction. *Human Brain Mapping,* 41: 1435–1444, 2020.

[53] Z. Zhang, Y. Genc, D. Wang, M. E. Ahsen, and X. Fan. Effect of AI explanations on human perceptions of patient-facing AI-powered healthcare systems. *J Med Syst* 45: 64, 2021.

[54] N. Luhmann. Trust and power: two works, 1st edn. Wiley, 1979. Chichester, New York.

[55] S. Meyer, P. Ward, J. Coveney, and W. Rogers. Trust in the health system: an analysis and extension of the social theories of Giddens and Luhmann. *Health Soc Rev,* 17: 177–186, 2008.

[56] M. Ryan. In AI we trust: ethics, artificial intelligence, and reliability. *Sci Eng Ethics,* 26: 2749–2767, 2020.

[57] S. M. West, M. Whittaker, and K. Crawford. Discriminating Systems: gender, race and power in AI. *AI Now Institute*; 2019.

[58] S. deBurca. The learning health care organization. *Int J Qual Health Care,* 2000; **12**(6): 457–8. doi: 10.1093/intqhc/12.6.457.

[59] IOM (Institute of Medicine). *Measuring the impact of Interprofessional Education on collaborative practice and patient outcomes.* Washington, DC: The National Academies Press; 2015. p. 182.

[60] T. Alqahtani, H. A. Badreldin, M. Alrashed, A. I. Alshaya, S. S. Alghamdi, K. bin Saleh, *et al.* The emergent role of Artificial Intelligence, natural learning processing, and large language models in higher education and research. *Res Social Administrative Pharm.* 2023 doi: 10.1016/j.sapharm.2023.05.016.

[61] D. Pinto dos Santos, D. Giese, S. Brodehl, S. H. Chon, W. Staab, R. Kleinert, et al. Medical students' attitude towards Artificial Intelligence: a multicentre survey. *Eur Radiol.* 2018; **29**(4): 1640–6. doi: 10.1007/s00330-018-5601-1.

[62] P. Gainza, S. Wehrle, A. Van Hall-Beauvais, A. Marchand, A. Scheck, Z. Harteveld, et al. De novo design of protein interactions with learned surface fingerprints. Nature. 2023; 617(7959): 176–84.

[63] K. Atz, F. Grisoni, and G. Schneider. Geometric deep learning on molecular representations. Nat Mach Intell. 2021; 3(12): 1023–32.

[64] M. E. Mock, S. Langmead, and A. C. Russell. AI can help to speed up drug discovery—but only if we give it the right data. Nature. 2023; 621(7979): 467–70.

[65] S. Warnat-Herresthal, H. Schultze, K. L. Shastry, S. Manamohan, S. Mukherjee, and V. Garg, et al. Swarm Learning for decentralized and confidential clinical machine learning. Nature. 2021; 594(7862): 265–70.

[66] J. Jimenez-Luna, F. Grisoni, N. Weskamp, and G. Schneider. Artificial intelligence in drug discovery: recent advances and future perspectives. Expert Opin Drug Discov. 2021; 16(9): 949–59.

[67] B. Van de Sande, J. S. Lee, E. Mutasa-Gottgens, B. Naughton, W. Bacon, J. Manning, et al. Applications of single-cell RNA sequencing in drug discovery and development. *Nat Rev Drug Discov.* 2023; 22(6): 496–520.

[68] P. Ma, H. M. Amemiya, L. L. He, S. J. Gandhi, R. Nicol, R. P. Bhattacharyya, *et al.* Bacterial droplet-based single-cell RNA-seq reveals antibiotic-associated heterogeneous cellular states. *Cell.* 2023; 186(4): 877–91.e14.

[69] M. C. R. Melo, J. Maasch, C. de la Fuente-Nunez. Accelerating antibiotic discovery through artificial intelligence. *Commun Biol.* 2021; 4(1): 1050.

[70] S. A. Alowais, *et al.* "Revolutionizing healthcare: the role of artificial intelligence in clinical practice." *BMC medical education* vol. 23,1 689. 22 Sep. 2023, doi:10.1186/s12909-023-04698-z.

Index